PROJECT MANAGEMENT FOR EXECUTIVES

and Those Who Want to Influence Executives

Steve Houseworth, PhD, PMP

iUniverse, Inc.
Bloomington

**Project Management for Executives
and Those Who Want to Influence Executives**

Copyright © 2011 by Steve Houseworth, PhD, PMP

All rights reserved. No part of this book may be used or reproduced by any means, graphic, electronic, or mechanical, including photocopying, recording, taping or by any information storage retrieval system without the written permission of the publisher except in the case of brief quotations embodied in critical articles and reviews.

iUniverse books may be ordered through booksellers or by contacting:

iUniverse
1663 Liberty Drive
Bloomington, IN 47403
www.iuniverse.com
1-800-Authors (1-800-288-4677)

Because of the dynamic nature of the Internet, any web addresses or links contained in this book may have changed since publication and may no longer be valid. The views expressed in this work are solely those of the author and do not necessarily reflect the views of the publisher, and the publisher hereby disclaims any responsibility for them.

Any people depicted in stock imagery provided by Thinkstock are models, and such images are being used for illustrative purposes only.

Certain stock imagery © Thinkstock.

ISBN: 978-1-4620-6158-7 (sc)
ISBN: 978-1-4620-6156-3 (e)

Library of Congress Control Number: 2011918829

Printed in the United States of America

iUniverse rev. date: 11/3/2011

Table of Contents

Introduction: Read Me First:	ix
Section I: Perspectives	**1**
Chapter 1: Basic Principles	1
Adding value / Delivering value	1
Providing what is needed to add value	2
Defining work	2
Prioritization	2
Time Exchange to Plan versus Fix	3
Planning is real work that provides control	4
Managing potential problems, changes and risks	4
Letting the Doers Do! Projects as an Entity.	5
Chapter 2: Contrasting Perspectives.	6
Demystifying Project Management	7
Everything reduces to Prioritization Perspective	9
Project Flow Perspective	11
Figure 4: Time Exchange from fixing problems at end to improved planning that reduces problems	12
Flow backward to reinforce the project perspective	15
A learned success versus learned mess perspective	16
Your primary role for projects	17
Chapter 3: Projects as an entity	18
Operational Definition of Project	18
Defining Success	20
Chapter 4: Characteristics of Successful Projects	22
The First Central Question: What do you want project management to do for you and your organization?	22
The second central question: What are your assumptions about project management?	24
Section 2: The Look of Success	**27**
Chapter 5: What a Successful Project Looks Like	27
Define Success	29
Chapter 6: Key Success Factors	33

Predicting and Controlling Success	33
A Top 20 List	33
Understanding the Success Factors	38
Chapter 7: Create Your Success Framework	**45**
Grab the Low-Hanging Fruit First	45
The top Five plus Change Management = Heroes	46
Two Important Role and Environment Adjustments	47
Chapter 8: Slogans and Handles	**51**
Overview	51
Handles	51
Project Management Slogans	52
Section 3: Facilitating Success	**57**
Chapter 9: Soft Skills Rule	**57**
The Impact of Facilitation	57
Why Soft Skills Rule!	59
Soft Skills Training	60
Ishikawa or Fishbone Diagramming	61
Structured Decision Analysis or Pugh Matrix	63
Chapter 10: Knowledge Area Topics	**66**
Overview	66
Scope management	66
Time management	68
People management (HR)	70
Quality management	71
Communications management	72
Risk management	73
Cost management	73
Procurement management	75
Integration management	76
Chapter 11: Managing by Metrics	**79**
Value of Metrics to Success	79
Standard Project Management Metrics	80
Custom Metrics	87
Chapter 12: Methodologies	**90**
Overview	90
Waterfall Methodology	90
Agile Methodology	92

Sprints	92
Closing Comments	**95**
Appendices	**97**
HELPFUL TOOLS	**97**
Top 20 Success Factor Table	97
Facilitation Tools	98
Iron Triangle and Three-Legged Stool Graphic	99
Index	**101**

List of Figures

Figure 1: An inside-out perspective can clarify confusion of project details — 7
Figure 2: Getting to the sweet spot of executives and project team perspectives — 9
Figure 3: Prioritization compared to perspective of executives and project teams — 10
Figure 5: Define the topic — 13
Figure 6: Iron Triangle and Three-Legged Stool — 21
Figure 7: Wasted Time is Wasted Money — 54
Figure 8: How Bad do you Want it? — 56
Figure 9: Fishbone diagram examining excessive heat — 62
Figure 10: Fishbone diagram for planning success — 63
Figure 11: Metrics Fence Example — 82
Figure 12: Waterfall methodology — 91

List of Tables

Table 1: Factors Defining Successful Projects — 32
Table 2: Top 20 Success Factor Table — 35
Table 3: Success Factor Rating With Bar Chart — 37
Table 4: Decision Analysis Matrix Chart — 65

Introduction: Read Me First:

This book is intended for two audiences:

First: Executives who want projects in their organization to succeed.

Second: People who want to influence executives to improve project success.

This book is the result of frustration. Not necessarily frustration with executives, rather frustration at experts who complained about executives. During my early project management years I heard a lot of "experts" and "consultants" complain and regale situations where executives unwittingly or knowingly hurt projects. But I heard no solutions. So I finally asked some consultants, how they influence executives to help rather than hurt. "Nothing" was the answer. Having a background in education, group processes and psychology I knew an opportunity existed to educate executives about how they – Okay you, could help projects succeed rather than impede success.

Some books seem to present project management as a panacea. I don't. It is a set of practices that only work as well as people apply them and structure work environments so the practices can work.

If all projects in your organization flow flawlessly, without challenges and complete within time and budget, and produce quality deliverables, then this book may be confirmation that your organization has great practices. However, if project execution in your organization has room for improvement, this book will help you understand both how projects work and how executives can support project success.

Towards this goal of helping projects succeed, this book contains ample content, exercises to apply, ideas for process improvement and maturing both the work environment and staff. Two quick examples:

- Understanding the perspectives of executives versus project teams
- The concept of a time exchange to encourage more planning and fewer problem fixes.

The book is written to be relatively short and concise. It is not written to be an exhaustive analysis or compilation of reference sources. Rather, I wanted this book to be:

- An easy but meaningful read
- A ready reference for practical application
- An idea generator.

I think you'll get value from picking it up often and rereading sections, exploring references and using the suggested active exercises to reinforce principles in your organization.

One last book note: Because the focus of this book is project management for executives, some work environment suggestions are included, but this really is a different topic that is addressed in my companion book <u>A Fresh Look at Improving Your Work Environment Using Project Management Principles</u>.

Section I: Perspectives

Chapter 1: Basic Principles

These basic principles are sprinkled throughout the book, but I feel they are so important that I wanted to present them right up front. This also makes finding and reviewing them quite easy. So, here you are.

Adding value / Delivering value

Project management was not ordained by God or fashioned from the big bang via subatomic particles. Project management is a completely contrived human endeavor that is intended to add or deliver value. This deserves emphasis because I know that project management has in many industries been a key buzzword for the past couple of decades. Some implement project management rigidly as in "This is THE only way to run projects." In other organizations project management has become ubiquitous; considered like part of the furniture. "Oh, just pull that convenient risk assessment unit over here while we talk about work." "No, no! Don't grab that change management unit. It has squeaky springs, is uncomfortable and people get wrankled over the decisions it forces us to make."

Project management as a completely contrived human endeavor has evolved and is intended to add value by completing work. No one way exists to implement it. In fact, if you can deliver value without it – Great! If your organization has difficulty delivering value, or meeting timelines, or fixing a lot of defects after delivery, etc. then project management implemented well can help.

As an executive you are in an ideal position to actively contribute to adding and delivering value.

Providing what is needed to add value

I like to emphasize that the component parts of project management are included <u>because they work</u>! As a completely contrived human endeavor, change management is included because it works, even if people get wrankled over the decisions it forces us to make.

Adding value requires certain ingredients and processes. As an executive, you are either willing to provide what is needed to add value or not. You really can't expect to add value through work that requires 5 FTE by providing only 1.5 FTE. I see companies and teams ensure adding value by emphasizing foundational processes like building agreement of the project scope, of defining work packages, of interactions and the need for detailed test plans or verification steps.

<u>As an executive, you are in an ideal position to provide what is needed to add value</u>.

Defining work

Defining work is a basic principle because this ensures everyone understands the work in the same way, i.e. the goals, objectives – everything about it. I present examples of agreement and of disconnects throughout the book because this is so important. I've seen the consequences of this in the form of success and conversely in the form of - as colleagues label it - "swirl".

Don't assume everyone understands things the same way, or even your way. Sometimes you may even need several weeks of a planned and coordinated campaign to explain what is intended, to show how it is valuable and build relationships so that others <u>want</u> to be spokespersons for your campaign.

<u>As an executive you need to ensure the work is defined</u>.

Prioritization

Everything really is a matter of prioritization. Delaying projects because you don't have the capital to invest is a legitimate prioritization. Building labs to conduct research and development is a prioritization over other uses for that money. Deciding to use formal, structured processes to control work, i.e. project management, is a prioritization decision. Providing only 1.5 FTE needed to complete the work requiring 5 FTE is a prioritization. You get the point.

As an executive you need to make prioritization decisions that support delivering value.

Time Exchange to Plan versus Fix

The concept of a time exchange to plan versus fix is one of the most important basic principles I have encountered in every work environment and every industry. This is the embodiment of the adage "never enough time to do things right, but always enough time to do them over again."

The time exchange is the "thought prior to action"; the "ounce of prevention versus the pound of cure." The more time spent up front figuring out how to do the work, how to prevent problems, how to sequence work steps, how to get others to be on your side, the less time is needed to fix problems later.

I developed the explanation of a time exchange after several situations where people said "we don't have time to wait and figure all this out", but they ultimately spent more time pausing the work to figure out what to do next than if they had taken an extra two weeks up front to figure it out. I began to present: "Think of this up front work as a time exchange in which you reduce swirl or time throughout the project by spending as much time as needed planning before you begin the work.

As an executive support the time and effort to figure things out up front, as a time exchange to prevent more time needed to fix problems later.

Planning is real work that provides control

As I just alluded, planning provides a level of prevention and a level of control. However, in many organizations planning seems to be considered not real work. "Why are they spending so much time in meetings, drawing diagrams and writing papers? When are they going to begin the work?"

Well, planning is real work. It is fundamental work that facilitates subsequent work.

Even organizations that spend time planning can have a tendency to plan only up to their psychological comfort level; sometimes not completing inevitable steps. I do like to emphasize that experienced project managers don't need to know a particular subject matter to realize that planning may not be complete or that risks may be inherent. This type of inquiry, discussion, and documentation is real work that both subject matter experts and those experienced in managing work need to do.

As an executive you need to actively support the basic principle that planning is real work.

Managing potential problems, changes and risks

Even the best planning can't anticipate all situations, but it can develop processes to handle situations. This is the basic principle. Throughout the book I present examples of how processes can lead to effective resolutions, even for situations not known earlier.

If your organization does not use processes to handle these type of situations, don't sweat! Project management includes these. Why? Because they work. In fact, a real and practical value of project management are the use of best practices for handling problematic situations like something changing or a potential risk actually happening.

As an executive you need to emphasize and insist that processes be developed and used to manage problems, changes and risks.

Letting the Doers Do! Projects as an Entity.

This basic principle is that projects are worked by dedicated teams whose mission is to complete the scope of work. Once the project is <u>defined</u>, <u>planned</u>, <u>funded</u>, etc. and have what is needed to be successful, the team needs to be supported but not interfered. I like to say that the project is now an "entity" that includes scope, processes, people, funding, etc. As such, let the doers do! One colleague refers to this as "self-directed teams." And I think this is a fairly accurate description. After all, they have everything needed to be successful or, if they don't at least they have processes to manage the things they do need.

Unfortunately, projects are often asked to add pieces of work, to do more with fewer people or money; to shorten testing, etc. These are all examples of change and some changes, like shortening testing, also introduce risks.

<u>As an executive you need to support the project entity by providing what it needs to be successful and let the doers do.</u>

Chapter 2: Contrasting Perspectives.

Whether your company is large or small, most of the work is probably organized as projects intended to enhance value or provide new value. Yet projects in many industries either don't complete or complete beyond the established time frame and cost much more than budgeted. This reduces the value the project represents.

Project management truly is all about the work. It is about completing work to add value to your organization. All of your company's processes, metrics, roles, controls, best practices, etc. are focused on one thing: completing work to add value to your organization. Theory and best practices around project management are the result of practical experiences - both good and bad. So with this type of foundation for projects, why should projects not complete and add the needed value to your organization? This book addresses some key reasons. For now, two reasons are:

1. Your understanding of how projects actually operate and
2. Your active involvement with projects.

Now you might ask: "While knowing project management intricacies would be an asset for helping me understand projects, are not project details the responsibility of a project manager? Why should I learn project management?" Excellent point! As an executive, you don't need to know project management intricacies. In fact, they can seem a bit mystifying; like Merlin the magician reciting Latin from his gigantic book of incantations and potions. Rather you need to be able to <u>recognize</u> aspects of projects, such as when someone is changing what the project is supposed to deliver so you can ask questions like: "why is this change happening?" "Who is pushing this or authorizing it?" Similarly, when reading or hearing a project status report you can <u>recognize</u> issues that indicate effective use of time, steps to ensure quality or, <u>recognize</u> a definite risk and be able to ask the team for the steps to mitigate it.

Demystifying Project Management

A key feature of this book is demystifying project management for you. Here is an example of how project management need not be a mystery. Notice the three times <u>recognize</u> is used in the preceding paragraph. Executives without a sufficient "inside out" perspective may not even recognize factors or situations that help or hurt projects. Just being able to recognize such factors will be another tool in your arsenal to add value for your organization.

I'll begin the demystifying process by presenting what I have seen as two somewhat different perspectives of projects by executives and project teams. Often an executive's view of projects is what I call from the "outside looking in." You sponsor work, it gets assigned, you want to monitor status and see results. Conversely, project details are from the "inside looking out" by the people doing the work.

Figure 1: An inside-out perspective can clarify confusion of project details

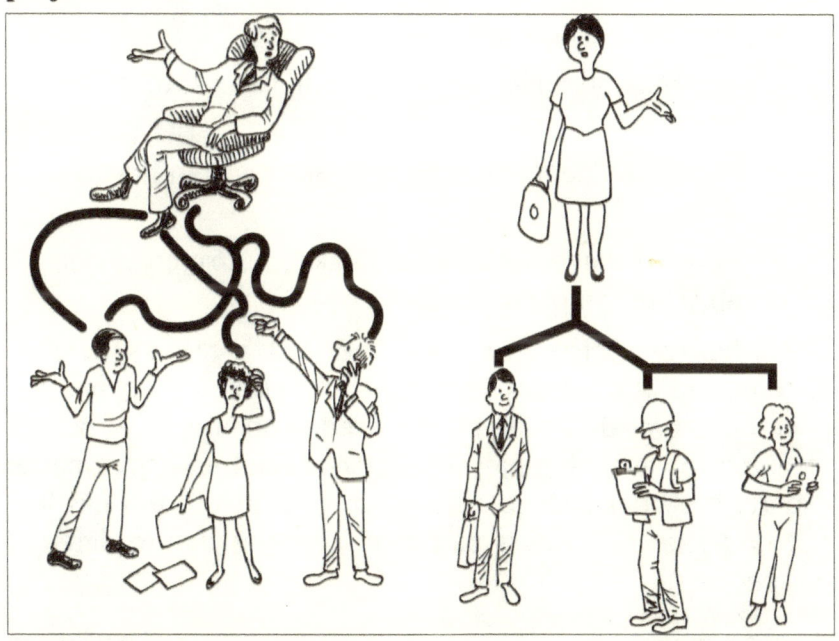

"From the outside, projects can seem unstructured and unclear. From the inside, projects are quite structured with clear paths."

Executives tend to be protective of factors like delivery dates and costs, i.e. when you want work completed and how much it costs. Project teams tend to be protective of their mission – the work they are assigned to accomplish and whether they have the time, people and other resources, and budget. These two perspectives can create substantial disconnects. I want to help you – as an executive - have that project "inside-out" perspective in order to remove this disconnect.

As an executive you probably focus on

- The organizational need for the work or deliverables.
- Value to be gained.
- Practical realities of how much time, cost and people you can afford to devote to the work.

In contrast, a project team perspective is that they have been given an assignment – a mission to complete. In order to complete that mission, they need to

- Analyze and plan sufficiently.
- Estimate the time, cost and people needed to do the work.
- Determine dependencies and risks that can hurt the work.
- Have a management champion, like you, to protect their ability to complete the work.

Often, these two perspectives create the need to reach a sweet spot or consensus of what can be delivered for what the organization is willing to provide. At this point, I need to point out that project planning and analysis represent reality – what the subject matter experts have determined is needed to complete the mission. Not understanding this is part of disconnects due to different perspectives.

Here is a graphic illustrating the two perspectives and getting to the "Sweet Spot".

Figure 2: Getting to the sweet spot of executives and project team perspectives

Executive Perspective	Executive Focus	Project Team Focus	Project Team Perspective
Organization Need	Here is what I want and what I can provide to do the work!	Here is what we need and how we'll do it!	Mission: To deliver the goals and objectives.
Goals and objectives			Detailed planning to know what is needed to succeed
Costs / Funding the org. is willing to provide			Having the people, $, resources, etc. to deliver the value
Time frame <u>desired</u>			Time frame <u>needed</u> to succeed
Receiving and using what the project delivers			Satisfaction of adding value to the organization
Sweet Spot — What projects can deliver for what is available.			

Everything reduces to Prioritization Perspective

Ultimately what projects need to be successful and what your organization provides to the project can be considered a matter of prioritization or importance. How important is the value you want compared to the investment needed to deliver the value?

Based on my experience project teams understand this quite well. In fact as I mentioned, the analysis and planning by projects is specifically to reflect reality then move to reaching the "sweet spot" by showing what can be delivered for what is provided. You may be able to get 70%, 80%, 90% or 100%. I can't assess how well you and other executives in your organization understand this. I simply want to present this as perspective for how to understand projects and view them from the "inside out" so you can get the most value for your organization.

Here is another chart highlighting the underlying issue of prioritization compared to the perspective of executives and projects. Note that these are some things that can happen, not an exhaustive list.

Figure 3: Prioritization compared to perspective of executives and project teams

Things that can happen	Executive Perspective	Project Perspective
Staff time is needed to do some other work. Sometimes called the "immediate crises".	This other work is important and needs to be done. Projects are longer term so, make some team members available to do it.	We need to evaluate the impact to our work schedule if you use team members for this other work. Are you willing to accept the impact to the project?
We want to add or remove some deliverables.	The project is funded and has the people. The changes seem logical and reasonable.	The project is planned for the original work. New work needs to be analyzed so the impact is known. What is priority?
Can't determine the project status.	Need to know if the project is on schedule, within budget, etc.	Ouch! We missed that aspect of planning. How important are these metrics at this point?
Project is way over or under budget.	Concern that planning is inadequate and need confidence that the work will be delivered. Need detailed explanations.	Planning was based on estimates. You are right, real costs should have been monitored more closely. The project will provide details then replan costs and time frames.
Project encounters something unforeseen / unknown.	How can this be resolved? ...or... Should this not have been known?	We still want to complete the work but, this is a new item that no one forsaw. It could affect costs, dates, etc. With this new issue, what work is still priority?

Have you experienced other similar issues that are a matter of prioritization? **As an active exercise**, take some paper and list situations then reflect on whether these are matters of prioritization?

As another example, one company I worked for readily accepted new work from its sales force with delivery dates that did not consider the company's capacity to complete the work. Evaluating which project was more important, staff vacation schedules, internal support work, etc. was not done. These would just have to work themselves out. The clear message from executives was, the sale/contract is the priority.

Project Flow Perspective

I've heard many people say that project management practices sound good in theory but do not work well in practice. Controls and processes just create paralysis by analysis, too much bureaucracy, etc. I've heard statements like:

"We need to complete work quickly, be agile – you know. We really don't have time to follow project management."

Well, there are two sides to this coin:

- Heads: Don't work carefully enough and costs may continue and may increase well after delivery; whether you deliver a road, software or widget.
- Tails: Add too many processes and your company will spend more time defining and documenting than building the road, software or widget.

I ask people to consider the time spent in planning as a <u>time exchange</u> in which time to fix problems at the end of a project is exchanged to earlier in the project to figure out how to prevent problems. In fact, measurements can be gathered to determine the value of catching a potential problem early rather than fix problems after delivery. This topic is covered in detail in my book "Creating a Project Management Work Environment."

Figure 4: Time Exchange from fixing problems at end to improved planning that reduces problems

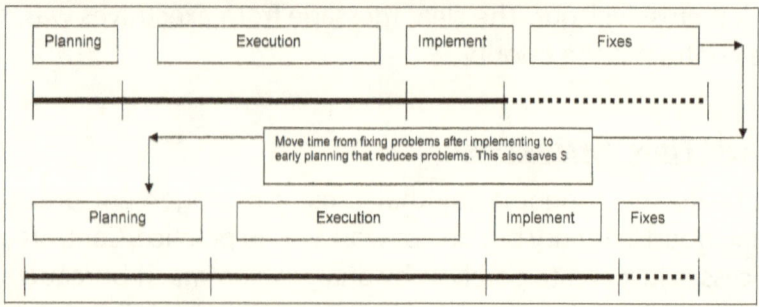

NASA faces risks that may justify the extraordinary time defining processes and documentation. Most other organizations do not face risks of NASA's magnitude. The answer is to have the proper balance for your organization. I also must say that I've seen project management rigor used quite effectively to add value and avoid both paralysis by analysis and the long tail of rework to fix defects. Said another way, project management is a facilitator rather than an inhibitor. This book will help you make project management a facilitator for adding value.

To this end the following set of steps shows my version of the ideal steps or phases of projects. Standard project terminology and some notes are included to provide a project vocabulary and rationale. Again rather than delve into details, this chart presents project flow appropriate for an executive level. As you work with projects in your organization you'll be able to recognize when issues during Execution and Control are due to inadequate detailed analysis. In these cases you can insist that the team perform the analysis and improve their project plans. In other cases you'll recognize that potential issues were prevented because you took the time to define and reach agreement about the project scope during the Initiation and Develop Project Charter phase.

Figure 5: Define the topic

Define, so everyone understands

Step 1: Determine what the organization wants the project to deliver. This is considered part of Initiating Projects and is a crucial foundation.

Step 2: Agreement about what the project should deliver, stated in a formal document. Get all supporting executives to sign the formal document.

Step 3: Acceptance of the project charter as the official mission.

Acceptance does two things. First, it documents that people agree so they can't say they did not know. Second, the project can now be viewed as an 'entity' that has a mission, leadership, a team, budget, etc.

All of the things needed to begin the project should be stated in the project charter.

Oh, Steps 1, 2 & 3 could take 5% to 10% of total project time. Initiation and agreement are that important. Remember that planning is <u>real</u> work.

Step 4: Detailed analysis of what is needed to complete the mission.

Steps 1-3 can be viewed as the initial impression or "blank" of a coin or initial sketch of artwork. Step 4, as the detailed analysis should take 10% to 20% of total project time. This is the thought before action; the opportunity to think through all aspects of completing the mission. As such, this requires a team of experts – both those who have experience doing the work and those who have experience implementing and supporting the deliverables after they are delivered. More about these in other sections of the book. Leave no stone unturned to determine all of the details.

Step 5: Acceptance of the project plan and schedule.

Acceptance recognizes agreement with what the team needs to complete the mission.

Also, plans should include handling of risks, changes, escalations to resolve situations, etc. <u>Similar to Steps 2 & 3 get all supporting executives to sign the document</u>. Oh, by the way…as an executive you'll be demonstrating your active involvement and your commitment to the project by a thorough evaluation of the plan and how promptly you sign it. This is also addressed in Chapter 6 as Key Success Factors, in which the top three success factors are active involvement of key management. Approving the project plan and schedule promptly demonstrates active involvement.

Step 6: Run the project.

This should take 60% to 70% of total project time.

This is where all of the plans are used to do the work, monitor progress, provide documentation, reports, etc.

Step 7: Formally accept project deliverables.

Deliverables can be completed at various times during the project. Such staggered deliverables should be included in the project schedule and planning documents.

Step 8: End the project

This should take about 5% of total project time. The phrase "the job isn't over until the paperwork is complete" should come to mind. All contracts, billing, hold-back payments, proof of performance and authorizations need to be completed.

Flow backward to reinforce the project perspective

I am a lifetime distance runner. Significant races, particularly marathons, require training to minimize risk in addition to ensuring fitness. The only thing worse than months of training and all of the discipline and pain management it entails, i.e. detailed planning, is to have miscalculated or missed some aspect then to get injured or 'hit the wall' somewhere during the 26 miles. To this end, I would plan my training from my current level of fitness to the race date. Just like listing the steps 1-8 above. Then to be sure I did not miscalculate, I would work backward from the race date starting with a rest period just prior to the race. Then, I would compare both training plans for differences and to make adjustments.

I want to use this same practice to examine these ideal work steps by working backward from the end of the project. I think by comparing the start-to-end flow with the end-to-start flow, you'll gain much perspective about projects and why they operate as they do.

So, take some time to start at Step 8, listing all of the things you expect to be complete and that your organization will need to do or have in place, e.g. authorization forms, at the conclusion of the project. Work backward step by step to evaluate what is needed to accept each project deliverable in Step 7; to execute each phase of the project in Step 6, to accept a plan that will complete work in Step 5, etc. Take notes and compare your list with activities your project teams develop. This can actually be a fun exercise fostering professionalism with your project teams and a greater sense of commitment from you to your project teams.

A learned success versus learned mess perspective

Now that we have covered perspectives from the "outside in" and "inside out" I want to present another perspective – that of learned success versus what I call "learned mess." This is a slight variation from the psychological "learned helplessness" phenomena.

You've probably heard this definition of insanity:

- Insanity = Doing the same things repeatedly and expecting different results.

Most people and organizations try to avoid working insanely. However, people and organizations can become accustomed to working without support or a lack of organizational process improvement or "no one is willing to take on the tough challenges." To that end, here is a variation of the insanity statement. Think about a term or label for it:

- Doing the same things repeatedly knowing you'll have problems. = what term?

You may wonder who would do this. Well, I've seen people who call themselves professionals do this project after project, year after year. In contrast, I've also seen professionals create success by using key success factors to avoid both insanity and "whatever-else-you-may-call-it".

Not having an ethic to ensure success by developing purposeful plans and by not rewarding people actively working to prevent problems can lead to a "learned mess" perspective. Ever hear statements like: "Nobody wants to escalate problems because that just creates resentment." Or "Executives listen but nothing ever happens." These are symptoms of a learned mess environment.

Development Exercise Note: By the way, a great staff building exercise is to contrast these two statements and discuss possible labels. If you conduct this exercise, don't stop with discussion. Actually extend the exercise to include action plans. Maybe posters with slogans can be posted. Real work examples showing how this statement was avoided can be gathered and discussed. Bookmark this page so you can return later to brainstorm other applications.

Your primary role for projects

As an executive, you are in perfect position to create success and add value to your organization. You are also in perfect position to recognize factors that create problems for your projects. Again, project managers know project processes in detail. <u>You need to recognize the key features of projects</u> so that you can partner with the project managers to ensure success.

Executives are not necessarily the primary source of changes, although they can be a significant source. I'll address sources of changes and how they can be controlled later. Rather, <u>executives like you are the primary source to protect projects from changes that reduce value</u>. Also, I understand that most executives do not realize they can both introduce change and protect work from change. In short, they are not familiar enough with how work is initiated, planned and executed to know how they are helping or hurting the work. Ahhh…the "inside-out" perspective.

Chapter 3: Projects as an entity

A necessary step in understanding projects is defining projects and success. After this, I'll link success with features of projects to complete the loop. This won't take long, I promise.

The term 'project' can be used colloquially or operationally. A colloquial use would be if you have two projects to complete around the house this weekend. You won't need to explain what you mean by the term. Fix a gate; clean gutters; throw away old margarine containers so you can regain cabinet space – some discrete or specific set of work.

Operational Definition of Project

Operationally in most industries and professions, 'project' also means some specific set of work with these two features: a definite start and definite end. Whether your organization considers these three components as the key features of projects, this is the operational definition used throughout the book. Let's take a real quick review of these components.

- Specific set of work: This is the project <u>scope</u>. The work is producing one or more 'somethings' that are often called deliverables or work products. This is important for formal project management because the specific scope leads to:
- Getting people with the right skills to complete the scope.
- Scheduling the work so you know how long it will take. People, schedules and other things equal money so,
- Any changes to scope along the way affects all of the things needed to complete the scope.

Now a key feature of this operational definition is that the work is well defined. If the work is not defined well, then it can morph many times, causing significant problems for both the people and work effort needed to complete it.

As an executive, you need to be sure the work is well defined or <u>run a separate planning project to define the work</u>. Only if everyone agrees on the work, that is if it is defined well enough for everyone, then how to work on it becomes straightforward. This is one reason project management emphasizes spending significant time determining scope and planning the work to complete scope.

- Definite start: This is usually called the 'initiation' process, which is important because those who want the work completed are saying "Take your marks"… "Get set"…"**GO**!!!" These people are usually called sponsors and the definite start of the initiation process is a formal beginning that should mean they are willing to provide a budget, make workers available, etc.

- Definite end: This is called 'close-out', which is when all contracts are completed, final payments are made, all workers are released from the project and, in many fields a support organization takes ownership of the deliverables. In the example of a new house being built, the new owners move into the house and begin making mortgage payments. They take delivery and don't expect construction crews to return for uncompleted work or to fix defects. After close out, the project no longer exists. There are no recalls, no second efforts, no redoes.

Notice that thus far I have not added anything new under the sun. Well, except for a formal structure. You see, a formal structure is important and I feel necessary. Does your organization use a formal structure for defining, initiating and completing work? If not, try running two separate projects: One with your current informal structure and the other with a formal structure. See which provides more control, prediction, value and efficiency.

Defining Success

How about defining "Success"? In most cases if the deliverables or work products meet specifications, the project is successful. However, there is another side to this coin. Your organization may not consider the project successful unless all the deliverables are produced, or you make a profit, or any number of other criteria. Well, what about if some deliverables were added after the start and this caused the project to complete with a financial loss? In the new home example, what if the new owners wanted a couple more closets and additional insulation? Both of these features will increase costs and extend completion time – if these changes are accepted.

This is where I really need to clarify the difference between the colloquial and operational definition of 'project'. In the professional world success equals completing the specific set of work – period! <u>Any time the specific set of work changes, all of the things needed to complete the new set of work also need to adjust accordingly</u>. This may mean adding more money and more time. Note that I'm not saying changes are bad or to be avoided. Only that once a project is <u>defined</u>, any changes alter the definition which creates the need to agree on a new definition, then analyze the impact of the change to funding, staffing, etc.

In the world of project management, projects can only control three things:

- <u>Scope</u> or the specific set of work. The definition.
- <u>Cost</u> or the budget needed to complete the scope. People, tools, equipment, contractors, etc. are all included in the cost.
- <u>Time</u> or the total number of hours and duration to complete the specific set of work.

Think of scope, cost and time as constraints that must be balanced. If any one of these three things change, then the other two also need to change. After all, if the set of work is larger than the budget and the team does not have the time, how is this project going to be successful?

Think of this like an equilateral triangle in which all three sides need to be the same length or, like a three-legged stool where if any one leg lengthens or shortens, the stool becomes less stable. But if all the legs adjust, then the stool remains stable. Here are a couple of graphics to illustrate how scope, time and cost need to be in synch.

Figure 6: Iron Triangle and Three-Legged Stool

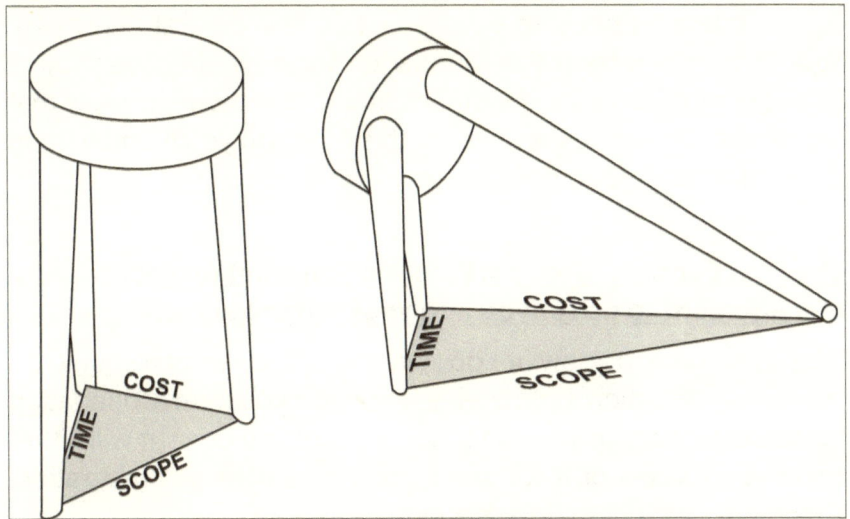

So, let's move on to what successful projects look like.

Chapter 4: Characteristics of Successful Projects

I want to prepare you for this chapter by examining your expectations. You see, your expectations will be a huge part of defining success within your organization and will continue developing that "inside-out" perspective.

To this interest I need to pose a first of two Central Questions. Your answer will form the foundation for how effective project management will be in your organization. Your answer is really that important. Oh, and by answering this question you'll understand project management much better.

The First Central Question: What do you want project management to do for you and your organization?

I don't presume to know nor do I want to tell you what it should do for your organization. I know what it can do if you form a foundation for success. Also, I know what it will not do if you don't. It will either bring you success or it will bring you contention and frustration. There is not a lot of middle ground.

But, I want to frame this question a bit more so that you understand the importance of the foundation formed by your answer.

A year after completing my PhD I worked for a sport-science publishing and education company. We developed books in the sport sciences and coaching education. We also conducted coaches training for the YMCA of the USA, several U.S. Olympic sport organizations, county and city park districts, etc. The youth sport coaching course used the philosophical foundation **"Kids First, Winning Second."** As I recall, this statement was presented right away - on the second slide of the two-day course and we referenced this foundation many times. The reason is that for most youth coaches the most prevalent role models are high-profile professional and collegiate coaches. However, these are not the most appropriate models for coaching children.

From the outset of the course, youth coaches needed to be directed to apply a youth-oriented coaching foundation. Coaching kids requires a focus on how youth sports can contribute to their development, i.e. developing skills, making friends and having fun, rather than winning games. We instilled this value in coaches so that they knew what they wanted youth sports to do for the kids. Oh, there were skeptics who said, "The college and pro model is just too powerful. Our coaches won't buy into this. Besides, parents want to see their kids win." But the program has been successful because of coaches who said, "You may be right, but I'm willing to give it a shot. We are in a great position to change parents attitudes."

Now is the time to begin forming your foundation for project management by answering this First Central Question. Remember, your answer forms the foundation for the level of success project management can add to your organization. Here are some thought-starters.

- Help manage by metrics.
- Provide schedules so we know when the work will finish.
- Provide management structure like having sponsors, PMs, team leads and worker bees.
- Quality needs to improve.
- Changes wreak havoc on projects so, we need a way to control changes.
- We have no consistency or standards. We want project management to help.
- Our clients require project management so, we use it to get their business. I don't see using it for our internal work. We need to improve time to market so we increase net profit.
- Provide a level of professionalism.

- Heck, I have no idea. I'm trying to learn what project management is all about so, I'm not sure I can even answer the question.

  Well, hang in with me. Answer as best you can. You can always revise it later.

Just as your answer to the First Central Question helps form the foundation for using project management, so also does recognizing assumptions about project management and evaluating whether they are valid. This is the Second Central Question:

The second central question: What are your assumptions about project management?

Your list of assumptions, perhaps presumptions might be more accurate, will be either reinforced or debunked during the remainder of the book. The exercise is extremely valuable because removing unsupported assumptions and reinforcing supported assumptions will combine with what you want project management to do, to form your project management foundation.

Here are some thought starters to help answer this question:

- Project management is complex and confusing, sometimes even contradictory. I see it as a necessary evil.
- Project management provides the structure, rigor and operational consistency needed to get positive results.
- Project management adds too much overhead and delays work, even for simple things.
- I push work through our pipeline and expect project management to do its thing. I don't want to know intricacies.
- I assume project management will generate metrics and status to allow me to evaluate the state of the work at any point in time and to predict with reasonable certainty what will be completed by when.

- I expect project management to build schedules so customers can see we have a work plan.
- I assume that everyone follows project management rules and processes.
- I expect project management to find ways to meet delivery schedules, not have people complain about things like work capacity, quality and changes.
- Project management is not responsive enough for our company. We need to be able to react quickly and move from start to finish faster than our competitors.

Section 2: The Look of Success

Chapter 5: What a Successful Project Looks Like

Playing mime games like charades with gestures of "looks like" are fun for social activities but they are absolutely frustrating when trying to know the status of work. I do want to emphasize "looks like" from the phrase "What a successful project looks like." Reports to read, charts and diagrams to analyze, metrics to compare are all <u>views</u> that should reflect projects from the inside out. <u>Successful projects will have planned and purposeful views</u> in the form of:

- Reports
- Charts
- Diagrams
- Metrics
- Architecture and designs
- Blueprints
- Schedules
- Network diagrams

I have seen organizations try to produce these type of documents well after the work is underway. This is <u>not</u> planned and purposeful. Rather, it is very much like trying to change tires on a car while driving on the freeway. Be careful not to add undue workload on people or interfere with project work, because this can spiral out of control. You'll always have time to ask about documentation and metrics during lessons learned reviews that can then be planned into the next project.

Make no mistake that these types of documents are detailed views into specific parts of those three project constraints of scope, time and cost that show whether your work is on track to succeed. If these types of documents are not part of the work in your organization, ask yourself

"Exactly how do I know what the work looks like?" "How can I support it if I don't know what it looks like?"

"How can I know if the project is on track for success?"

"What documents are available for me to see the status of projects?"

Same conclusion if you think producing these take too much time away from doing actual work. <u>Documentation of project status is actual work and should be factored into the project schedule</u>.

Even though you may have regular meetings with those doing the work, you won't get specific details unless they are using some form of these documents. The best you'll get is anecdotal information like:

"The team is making good progress."

"We're working through the database changes."

"The conduit is on site and being installed."

Unfortunately, these types of comments are not particularly accurate or meaningful views into the work. They really don't show what a project looks like. How much progress? How much longer to install the conduit?

"The finishing crew is scheduled for next week so will the conduit be hung and wires pulled by then or will I need to eat a day or two of costs until it is ready? Did our budget anticipate and plan for any delays?"

You should be able to see status - know what your projects look like. One motto I learned long ago: <u>Successful people can point to particular and specific factors that led to success, while those who</u>

<u>are unsuccessful usually point to many generalized factors.</u> Same is true of sport teams, businesses, landing a man on the moon, etc.

Are these the type of things you hear about projects?

"Well, you know, this is the first time we did this type of work."

"A lot of complexity with a lot of coordination that fell through the cracks."

"It was a good idea but not all of the details were worked out."

Hmmm. Sounds like what a great friend and running buddy says about the Cubs. When the Cubs are playing well Chuck can list batting averages, bases stolen, errors, home and away stats, but when they aren't playing well he generalizes with 'They've fallen on hard times.' No details. We've talked about it and Chuck agreed he does this. We are still friends and he still loves the Cubs.

<u>Successful projects have specific details that can be looked at!</u> Pardon the poor grammar.

Take an inventory of projects in your organization.

- What view do you have?
- Can you identify the specific factors that indicate success? That indicate whether you'll get the value you want?
- What would you like to see that you don't? This may be an unfair question if you aren't knowledgeable about project management. Before you finish the book you'll know things you'll want to see.

Define Success

During graduate school the teacher of a research methodology course invited a professor from the philosophy department as guest lecturer for the weekly two-hour session. I thought this was adding insult to injury – not much practical to come from any of this. But, the philosophy professor actually made sense and provided one of the best practical statements I've ever heard.

The topic at the time was measurement of things that are not tangible. Examples include psychological factors like aggression and affection. Social and business factors like loyalty and confidence. Hmmm…maybe even a factor like success. Well, the philosophy professor knew how to cut to the chase. He said:

"If you can define it, you can measure it."

I thought "that is a fairly elegant statement." The more I thought about it the more I realized that the definition of something became the pathway to its measurement. Hmmm. Same for defining project scope. If it is defined, then it can be planned, worked and completed.

So, let's define a successful project. Well, projects complete work, add value, have cost-benefit ratios, produce deliverables or products, involve people with skills, etc. A lot of people in project management paraphrase all of this into the phrase "complete work on time and on budget." Ahhh, if only success could be that simple. Reality is that change happens, schedules are interrupted, key people get sick or go on vacations, materials can be defective, weather may not cooperate. Sound familiar?

But even if success equals completing work on time and on budget, this raises the reality that you need to know details that lead to success and how the budget looks all along the way.

This leads us to the reality that any definition of successful projects needs to include some details about the <u>processes</u> involved in completing work. These processes are the equivalent of the individual steps in a 1,000 mile journey. Here are some examples:

- Regarding the planning process: The planning is complete. Sally has an established relationship with our supplier so, we know manufacturing durations of our vendors and order lead times. Fred in logistics has determined shipping details so, we know on-site requirements, like loading dock heights and, that the streets are all wide enough for trucks to reach the site.

- Oops. We haven't figured out on-site security and if we'll self-insure. I guess planning isn't complete yet.

- Regarding cost and budget processes: We need to monitor costs for two reasons. First, to know if the work is within the budget at each stage or month-by-month. Second, to know if any of the 10% contingency fund can be returned.

- Regarding communication processes: A lot of investors and city residents are counting on this new bridge. They'll want to know on a regular basis if progress matches the schedule. Local and national media will want to report and interview the crew. In order to present accurate and consistent reports that can't be misinterpreted, we should produce regular press briefings and train the crew that they don't give interviews. That's the job of PR.

You get the idea. Now, even if your projects are much smaller than building a bridge, defining success needs to include these processes. However, this does not need to become a laundry list.

Active Exercise: I encourage you to use the following chart to develop a definition of success for your projects. Start with factors you think will provide the best value for the effort. Treat this like a brainstorming session. Pull in colleagues to work through it with you. You'll probably need more than one pass to complete it. I seeded it with a couple of items to help you get started, but feel free to edit these and add your own.

Project Management for Executives

Table 1: Factors Defining Successful Projects

Defining Successful Projects			
Key Factors	**Why Needed**	**Measurements**	**What level does this exist now (1 = low to 5= high)**
Budget	Manage costs;	Monthly cost reports. Detailed budget expense items. Projections of costs for the next month.	3 – Receipts and invoices are easy. No format exists from accounting to track expenses or automate projections.
Work Schedules	To know the order work needs to proceed. To plan adjustments if work schedules slip. To know when staff with specific skills are needed; both ours and vendors.	Work breakdown structures. Network diagrams. Work dependency relationships. Vendor availability schedules.	2 – We use a spreadsheet to note some items. Mostly, the team has a general idea of what to do when.
Prioritization	A lot of dynamics can cause solving the immediate crisis to divert people from project work. Seems like a higher-level chief can slide in their work. This creates uncertainty with the staff.	Formal agreement among executives ranking the projects. Formal allocation of staff to certain projects. Formal agreement to allow project managers to 'own' projects so high-level chiefs can't interfere.	5 – Past experience led to management formalizing these last year.
Others			

Chapter 6: Key Success Factors

Most of us expect measurement in the form of statistics, some numerical values. Measurements can be quantitative using numbers and qualitative using checklists and descriptive labels. At this point I'm using qualitative measures. While numerical measurements are part of formal project management, delving into them at this point would complicate our vision of success.

Predicting and Controlling Success

Defining what successful projects look like not only provides measurement, it also allows prediction and control. Think about it. Determining key success factors allows you, as an executive, to instill these factors within your work environment. So in addition to monitoring the success factors throughout a project, you should be able to predict and control degrees of success. This gets back to the "sounds good in theory but doesn't work in practice" contention. Using key success factors can create successful reality. As a good friend is prone to say: "The best way to know the future is to CREATE the future!" As an executive, you are in perfect position to create success and show others what successful projects look like.

A Top 20 List

Another useful view of successful projects is the following list of my top 20 success factors. It is also included in the Helpful Tools section. This table uses a Likert scale that helps us see a probable, even predictable, level of success. When developing this table as one view of successful projects, I avoided a lengthy laundry list but I also did not have a preconceived number of factors. I'm very sensitive about stating that a certain number of anything is somehow official. Certainly no formal research led to this list. Ten factors were far too few. Stopping at 17 seemed too artificial. These 20 factors include

Project Management for Executives

all essential success factors and introduces other valuable factors that are explained in more detail later.

Become familiar with this, then review it again using the explanation following the chart and the success factor explanations in the next section.

Table 2: Top 20 Success Factor Table

Success Factors	To What Extent are these Success Factors In Place? Low to None = 0 or 1 Somewhat or sometimes = 2 Mostly but could be better = 3 Yes or appropriately = 4
1. Executive sponsor engaged.	0
2. Immediate sponsor engaged.	1
3. Project manager engaged.	2
4. Client / customer engaged.	3
5. Project control framework accepted and used.	4
6. Formal charter stating the project mission..	3
7. Deliverables are defined.	2
8. Stakeholders established	1
9. Management plans developed and authorized.	2
10. Core team established and allocated appropriately.	3
11. Project team has autonomy.	4
12. Reporting and metrics established.	3
13. Ethic of adding value to the organization in place (quality, changes, risks, issues).	2
14. Escalation paths defined & encouraged.	1
15. Progression of: scope to req. to work to schedule followed.	1
16. Contracts, SOWS, NDAs, etc. in place.	2
17. Change management established	3
18. Risk management established	4
19. Methodology to guide work in place.	3
20. Soft Skills / facilitation processes in place.	2
Score	45 = Questionable
Success Rating	Score = 0 – 29 Bad (Are you prepared for the cost of chaos?) Score = 30 – 49 Questionable (What is the cost of getting to 'Better'?) Score = 50 – 69 Better (Issues are probably manageable.) Score 70 – 80 Best (Architected for success.)

I seeded these with values and calculated a score so you could see a probable level of success. Simply determine the level for each factor then total the score. A project with these ratings and a total score of 45 is rated as Questionable for success; with the thought of "What is the cost of getting to a rating of Better?"

Staying with a visual theme, here is the chart implemented in a spreadsheet with a bar chart indicating the ratings. Pretty easy to see what is in place and what can be improved.

and Those Who Want to Influence Executives

Table 3: Success Factor Rating With Bar Chart

Success Factors	Rating: 1=Low to 4=High
Executive sponsor engaged.	0
Immediate sponsor engaged.	1
Project manager engaged as prime authority.	2
Client / customer engaged.	3
Project control framework accepted and used.	4
Formal charter stating the project mission..	3
Deliverables are defined.	2
Stakeholders established	1
Management plans developed and authorized.	2
Core team established and allocated appropriately.	3
Project team has autonomy.	4
Reporting and metrics established.	3
the organization in place (quality, changes, risks, issues).	2
Escalation paths defined & encouraged.	1
Progression of: scope to req. to work to schedule followed.	2
Contracts, SOWS, NDAs, etc. in place.	3
Change management established	4
Risk management established	3
Methodology to guide work in place.	2
Soft Skills / facilitation processes in place.	1

Understanding the Success Factors

Many of the success factors are self-explanatory, but explaining them will help provide focus for your vision of what successful projects look like and is a great orientation of project management presented later. Note: This is the most detail-concentrated portion of the book. Take your time and take a couple of breaks if needed to keep your eyes from glazing over. Also, this is a good section for continued reference to check on "what did Houseworth write about those success factors?"

Success Factor Explanations

1. Executive sponsor engaged.

Note: This takes time and commitment. Engaged = active participation. Most work, i.e. projects, reflect organization goals, objectives, priorities, portfolio management, etc. so they are either generated from an executive level or are approved at an executive level. Executives, like you, need to be engaged to do everything possible to ensure success. No, you don't micromanage. No, you don't run the project; that is the project manager's job. Yes, you do make sure these key success factors are in place. You do make sure measurements are gathered, reported and used. You do make sure that organizational policies, methodologies, etc. are followed. You'll need to report status of projects you sponsor so, you need to be engaged enough to set these expectations, provide the framework for success and hold immediate sponsors and project managers accountable. This is all part of #5 Project Control Framework Accepted and Used.

2. Immediate sponsor engaged.

Note: This takes time and commitment. Engaged = active participation. The immediate sponsor is often a manager of a team or area which will own the work or has the greatest vested interest. This area may be responsible for continued support, repairs, responding to issues from a call center, etc. This sponsor is engaged similar to an executive sponsor with more practical and specific issues like staffing and budget adjustments. This is the sponsor to whom the PM is responsible and accountable. PMs should have the authority to resolve project issues, but if the PM needs sponsor support then some issues and decisions may be escalated to this sponsor.

3. Project manager engaged as prime authority.

Note: This takes time and commitment. Engaged = active participation. Project managers manage the project. They should be responsible for the project team and vendors doing their jobs with quality. Similar to sponsors, PMs do not micromanage but they do pull together the project team to plan the work, then execute the work plan. If sponsors do their jobs and are engaged appropriately, PMs should have the authority and all resources to manage the work so that value is added to the organization. However, sometimes project managers need to manage multiple projects. A PM's degree of engagement should be appropriate for a project. Some projects may only require 25% commitment while others may require 100%. PMs should not be overcommitted. Over commitment is one way this success factor can be rated low.

4. Client / customer engaged.

Note: This takes time and commitment. Engaged = active participation. Because clients or customers receive deliverables, they need to be engaged. They are responsible for fundamentals like what they really want, approving designs and materials, reviewing pilots or mockups, testing or approving tests, helping resolve issues and making decisions. While this may seem obvious, the reality is that clients demonstrate their commitment and priority to a project by their level of engagement. Availability for meetings and completing assignments are ways customers show their level of engagement. Active engagement is a sure sign of probable success while low engagement is a sure sign of probable problems.

5. Project control framework accepted and used.

A project control framework consists of several success factors, including 1-4 above. <u>Most important is the agreement or acceptance to have a framework</u>. Organizations who consider a framework as a facilitator for success and have instilled this framework within the work environment are well-positioned for success. For example, dates are not committed unless the scope and requirements are known and approved. Many projects start marching toward a date not knowing if it is reasonable. A framework structures things that are necessary, sufficient and acceptable or not acceptable. Like a rowing team, everyone will be pulling together in synchrony. Picture a rowing team where one rower is out of synch and you can understand the equivalent of the additional water resistance on a project team.

6. Formal charter stating the project mission.

The project charter is the single most important document for a project and should be an essential component within an accepted project framework. Charters state 'what' the project is about but stop short of stating 'how' the work will be completed. The charter outlines the project and management expectations. Things like: the essential mission, features and components of the project, how it fits organizational or client goals, the envisioned deliverables, desired time frames, etc. The most effective charters I've seen are clear enough to be understood by just about anyone in your organization. Charters should be approved by all sponsors and clients/customers, ensuring that they are in agreement and that they all will support the project team to complete the work as it is defined.

7. Deliverables are defined.

Knowing exactly what is to be produced or delivered, i.e. deliverables, provide clear objectives and direction. This amounts to doing the right work in the right way. The degree to which deliverables are not defined creates ambiguity and contention. Often projects need to spend ample time to define initial visions of deliverables. Even these refined definitions can be considered deliverables, as in architectural designs, functional and process-flow diagrams. As long as the process of defining deliverables is included as part of the project work, leading to detailed deliverable specifications, this factor can be rated high. I call this working to know, then working what you know. If not, this factor should be rated low.

8. Stakeholders established

Stakeholders are those people or areas that are really impacted by the project. Stakeholders have a vested interest in the work and deliverables. As such stakeholders need to be known so the PM and sponsors can work with them appropriately.

9. Management plans developed and authorized.

Management plans are part of the project control framework. A separate management plan should be developed for each of the 9 PMI project management knowledge areas that apply to a project. Not all 9 knowledge areas are essential to every project. For example, some projects do not purchase goods or services or rent anything so procurement management would not be needed. However for those that are needed, authorization is essential to ensure agreement among sponsors, project managers and customers. One great example is stating how to recognize risks, rate risks and manage risks. Pretending risks do not exist or not stating them because of political factors is a core threat to project success.

10. Core team established and allocated appropriately.

A core team should be established, usually composed of key client staff, team leads, coordinators, supervisors and subject matter experts. Identifying members of the core team reflects clarity of the project mission, deliverables and impact to the organization. For example, projects may need a dedicated communication role and PR point person due to the impact to the organization, stakeholders, shareholders, etc.

11. Project team has autonomy.

The above roles, operational framework, documentation and plans provide project teams – which includes the PM – clear and specific direction to conduct the work and provide value to the organization. Teams also need the autonomy or independence to do the work. <u>This autonomy should be part of the project control framework!</u> Understand now why formal plans and approvals are key success factors?! The converse of this autonomy is actually micromanagement by others who aren't as close to the work as team members.

12. Reporting and metrics established.

Measuring project progress and ultimately success should be determined during project planning. This takes time and may be novel in your organization, but <u>you really need to determine what needs to be measured and how it should be reported</u>. This really provides efficiencies and the ability to make adjustments from the start. Trying to gather metrics to report progress part-way through a project is just plain difficult, takes time away from completing deliverables and usually provides incomplete information that might not lead to effective decisions. Specific project metrics are presented in Chapter 11 Managing by Metrics.

13. Ethic of adding value to the organization (quality, changes, risks, issues).

Because projects are work that add value to the organization, everyone should operate this way. This includes operating within the formal project control framework. Example: Formal change management should be used if something changes during a project. Even vendors should be encouraged and expected to raise issues if they see something that conflicts with adding value.

14. Escalation paths defined & encouraged.

Escalation of issues, clarification of scope, risks to meeting the schedule, budget increases, etc. are normal parts of projects. Knowing the escalation paths and that escalations are encouraged increases the probability of success. Not defining or not using escalation paths really increases the probability of contentions and decreases success.

15. Progression of: scope to requirements to work to schedule followed.

This progression is the appropriate way to determine what work the team is assigned, when it is scheduled and how the work can be measured. A controlled progression of defining scope, translating the scope to detailed requirements, then determining the specific work activities and sequencing within a schedule truly facilitates success.

16. Contracts, SOWS, NDAs, etc. in place.

Contracts, Statements of Work, Nondisclosure Agreements, etc. are both work facilitators and organizational protections. These documents ensure working relationships among entities such as vendors and clients, specify who does what for whom, liabilities, etc. Agreement to include these and not progress until they are approved should be part of the project control framework in #5.

17. Change management established

Changes occur on every project and are arguably the single most influential aspect of projects. Whether changes are small or large, a formal process to analyze and approve changes keeps everyone honest, informed and in agreement that the changes add value to the organization.

18. Risk management established

By definition risks increase the likelihood of problems. Identifying potential risks allows projects to develop both strategies and actions to either eliminate the risk, decrease the possibility of it occurring or reduce the impact if it does occur. A risk management plan increases the probability of success.

19. Methodology to guide work in place.

Methodologies are specific ways to control work. Methodologies specify work stages, criteria for moving from one stage to the next, criteria for ensuring quality, testing, etc. Methodologies can depend on industries and subject matter, but an appropriate methodology facilitates all aspects of projects.

20. Soft Skills / facilitation processes in place.

Most project work that leads to definitions, decisions, assessments, etc. involve what many people call soft skills and facilitation processes. The value of conducting efficient and effective meetings can free up hundreds of hours and surface many important issues during the life of a project. Situation analysis and decision analysis can complete in a couple of hours what may take weeks or months without formal facilitation processes. I consider soft skills so important that I devote all of Chapter 9 to this topic.

Chapter 7: Create Your Success Framework

Grab the Low-Hanging Fruit First

Success factor item 5 from Table 5, the Project Control Framework, is the focus of this chapter. Unless your organization has a mature project management work environment, creating your success framework will probably be an evolutionary process rather than an revolutionary process. I've seen both evolutionary and revolutionary attempted with the result that change becomes evolutionary regardless. This is primarily because project management actually entails work culture change that impacts:

- How people approach work. Oh, this includes executives.
- How work is packaged.
- Documentation, forms and tools.
- Staff training and maturation working within set processes.
- Organizational acceptance across business operations and enablement operations such as Information Technology, Logistics, Quality Control, etc.

I can't predict your opportunities or challenges, but I can say that you should target initially factors that provide the best value for the level of effort and your current work culture, i.e. the proverbial low-hanging fruit. Several reasons for this:

First, the low-hanging fruit are probably factors that your organization faces repeatedly or that creates particular pain points. Addressing these is a great way to show how formal processes can be facilitators rather than inhibitors.

Second, resolving some key problem factors presents the

opportunity to gather measurements that can lead to planning more measurements. For example, track the number of overtime hours logged prior to the change and for a few months after the change. If staff typically don't log overtime then begin tracking whether staff are staying late or coming in on weekends as much as they used to.

Development Exercise Note: Meet with other executives, managers, team leads or supervisors to brainstorm measurements in your organization that can show the value of adding a project control framework. Be sure gathering these measurements is simple and does not increase administrative overhead for your staff.

Third, this helps create the opportunity for teams and unit managers to experience working smart first, hard second. My experience is that once they get a taste of the benefits and know that executives support it, you'll have teams eager and willing to get to the next level.

Fourth, this begins to establish the various roles and responsibilities. Notice I say "begins" because some roles – like executives - will need to relinquish control while others will need to accept more responsibility. Achieving success in a few factors will generate confidence to do both.

The top Five plus Change Management = Heroes

I do recommend starting with the top 5 key success factors, plus change management, which is #17. If the primary roles for sponsoring, and managing work in your organization can accept working within a control framework and emphasize formal change control, you'll be heroes in a short time and…most of the subsequent factors will be implemented also. Select the next project to implement these. Gain some experience. Then, extend these to the next two projects. Gain more experience. By that time, you'll be ready to add much more formal project management, implement metrics, etc.

I can't say that these will solve a certain percent of your project work issues. I can say that agreement among the primary decision-makers both solves and prevents a lot of problems. In fact, the four roles of executive sponsor, immediate sponsor, project manager and client/customer should always be the people to resolve issues about how to add value to the organization. After all, it is their work.

Now to formal change management. Usually, whoever is the biggest chief; wields the biggest club - select whichever metaphor you want – ultimately wins changing aspects of projects. Formal change management removes the club, instituting a formal analysis of the costs, benefits, impact to schedule, etc. Basically, what is needed to incorporate the change.

Question: Who gets to decide?

Answer: A change committee, not just one person.

Question: What is involved in evaluating a change request?

Answer: Everything that is pertinent. The team looks at the amount of work, maybe even rework. Evaluates the hours and duration needed. Compares the cost to the remaining budget, etc. Sometimes a change results in the team figuring out that only some of the work can be completed unless the budget or time also changes, i.e. the Iron Triangle and Three-Legged Stool.

If you want to explore this more fully now, jump to Chapter 10: Knowledge Area Topics and get my other book <u>Creating a Value Added Work Environment Using Project Management Principles</u>. We strayed from creating your success framework. I need to get back on track!

Two Important Role and Environment Adjustments

Changing any part of a work environment poses challenges, particularly adding new processes and adjusting role responsibilities. I've found these next three factors to be significant considerations for creating a success framework that supports project management.

- **First: Project Managers can only do what your organization supports.**

If you as an executive and your organization support project managers, they can and will complete your work with quality. If you don't support them, then someone else – maybe many others – will play parts of a project manager's role. Leadership and decisions will fluctuate and the foundation projects need to be successful will not exist. This is why I included as a key success factor #11 Project Team has Autonomy.

The project manager needs to be supported within your organization, including having the autonomy to manage the project <u>based on the project charter</u>. Rather than other executives or clients telling the project what to do or changing scope, they should be asking the project what is possible. The project will evaluate requests then provide meaningful answers based on facts rather than guesses.

Now, this raises the important item of selecting project managers. Just like all roles and people, PM's personalities, work habits, ability to speak, write, develop rapport, etc. will influence their effectiveness. Very "blah" to neutral personalities can be quite successful while very effervescent and endearing personalities can fail. I look for people who are a "total package." I recommend highly that your organization develop criteria for becoming a project manager or to be used to hire project managers. Being blunt, you want to be in a position to support project managers that others already respect rather than people others consider inept or not trustworthy.

- **Second: Projects reflect reality.**

This returns to the "sounds good in theory" point. Conducted properly, project teams analyze all the factors needed to complete work, then develop plans to complete the work. Finally, the plans are executed and the work is completed. You are already familiar with this. The principles of project management are <u>very simple</u>! Don't be fooled into thinking that project management is complex. Similarly though, the analysis, the plans, the executing of the plans…these are all based in reality. Changes to that reality, without adequate analysis and planning is simply unreality. A cynical perspective might label it fantasy. Directing projects to do things that are not possible or to change scope several times is simply a formula for frustration, delays, crises and additional work to fix defects or add functionality after end dates.

Now, whether teams analyze work sufficiently and accurately is another matter. I want to share two stories about this.

On one project my team members were from the sponsor's unit. Every couple of weeks they reported that the area's subject matter expert (SME), who wrote the project charter, was

- Changing the projects assumptions,
- Changing position based on their analysis and
- Changing work scope estimates.

I, the SME and sponsor finally needed to decide what could and could not be accomplished. Their perspective was that the team members were not very good and did not understand the scope so, I as the PM needed to get them back on track. During the discussion I was able to explain specifics that caused the team members' concerns. The sponsor finally understood and we had the opportunity to surface that the SME only had two weeks to investigate and write the project charter, which made it not as complete and accurate as it needed to be. This was not nearly enough time.

Restated: The project was not defined clearly. The SME needed more time to develop the charter so that it reflected reality better before it came to my team. Result: the project scope was expanded, the end date did not change, but we were able to prioritize deliverables and complete the several most important by the end date.

Another project I inherited as a sponsor was initially estimated to take 1,500 hours and was thought to be on track. Oh, the PM also left the organization at the same time I inherited it as sponsor. The new PM needed two weeks to assess project status because the previous PM had not produced adequate project status measurements. Uh, Oh! The project was behind and almost through the 1,500 hours with a lot of work remaining. The business partners were great and continued to support the project based on cost-benefit analysis, even though it grew to 15,000 hours – a ten fold increase! Of course the end date also needed to be extended. My Director was not pleased because he promised a Vice-President that the project would be completed on time. Our conversation was not pleasant and included me chastising him for promising a date rather than requesting a date from the project team. He used the "that sounds good in theory…" line. I reminded him that the project team was reflecting reality and that the project scope had not been estimated accurately in the first place. Restated: The project team needed autonomy.

Projects do indeed reflect the reality needed to complete a discrete set of work, even if other people don't believe it.

Chapter 8: Slogans and Handles

Overview

Slogans and handles are great ways to encapsulate key success factors as phrases and mental pictures that help people see what successful projects look like. Certainly these are not exhaustive and not all are original "Houseworthisms."

Development Exercise: A great exercise is to write the handles and slogans on slips of paper, distribute them to people for discussion and their own descriptions. Then, bring participants back together to compare their descriptions with the descriptions presented here.

The exercise surfaces biases, pain points, and sentiments like "if we could only do this" and "yeah, like that will work in this place." Be prepared for these. I recommend conducting these exercises with upper management first to build support, then talking to unit managers and teams next.

Handles

Most of us know what slogans are but in case you aren't familiar with 'handles', they are short statements that people can visualize, grasp and use. To say "That sounds like a Dilbert" is to use the concepts of problems in the corporate workplace portrayed in the Dilbert cartoons. This is a 'handle'.

Other handles include the two statements from the Introduction, i.e. Insanity and that other term you assigned to "Doing the same things repeatedly knowing you'll have problems." Others:

- Like trying to hold a wet bar of soap in the shower. Squeeze too hard and it squirts out. Hold too loose and it slips out. Need to apply just the right pressure or firmness.

- Projects are about adding value.
- Are we adding value or reducing value?

Project Management Slogans

Plan the Work then, Work the Plan

Explanation: You've probably heard the Lewis Carroll saying that "If you don't know where you are going, any road will get you there." In project management terms, all work should have a plan to accomplish it. The plan includes how to make sure the work is defined, i.e. everyone working on it understands the work in the same way. The plan includes how to measure success, how to communicate the work status to people who need to know, how to know who needs to know, etc. If the work is planned for all the factors and contingencies, it can be executed, communicated and delivered without many difficulties. So, the level of planning should match the planning required for the type of work. The planning itself can take weeks, or months. As a general rule, about 5% - 15% of total project time. Another key point is that the work should not begin until the plan is complete enough to guide the work.

Get the Right People, at the Right Time, to do the Right Work, in the Right Way

Explanation: I think this slogan along with "Plan the work, then work the plan," encapsulates the essential features of formal project management. This slogan drips experience and reality. Think about the components:

- The right people are only known if the work is defined well enough to know the skills and numbers of the work force needed to complete it.
- The right time implies work sequences. Some work is dependent on other work. Also, that the amount of work time needed can be estimated and calculated so that people have sufficient time to complete it.

- The right work implies that the work is defined to a sufficient degree to know who, how and when it can be completed.
- The right way implies that the work has been analyzed, is planned and will proceed intelligently.

Work to Know, then Work What You Know

Explanation: Completing work is often a series of discoveries. Some are small while others are large. Formal project management follows the premise that what is not known can't be planned, scheduled or completed. So, this slogan emphasizes resolving what is not known so that it can be worked. Assumptions should be avoided and replaced with knowledge. The French attempt to build the Panama canal failed because of many unknown and unplanned factors – but people were busy. The U.S. canal work almost suffered the same fate until the chief engineer solved many problems with a solution of elevated locks, but only because he worked to know then planned to solve what was known. You probably have your own stories.

Work Smart First, Hard Second

The work culture in many organizations emphasize being busy as in busy-ness rather than effectiveness. Project management, in essence, emphasizes working smart rather than just working. I like to say that if we work smart first we won't need to work as hard later. This requires planning and solid execution. Even so, the ethic that busy-ness equates to meaningful work is tough to change in most work environments. Changing this requires at least these two factors: First, analyzing whether people are busy working smart or busy fixing issues because they have not been encouraged to work smart. Second, a cultural acceptance that thinking, planning, analyzing and documenting, etc. is real work. I've heard many executives say things akin to "Why do those people spend so much time in meetings, drawing models and talking? They should be at their desks or on the site working." As an executive you are in an excellent position to influence this change. A corollary to this slogan is presented next.

Project Management for Executives

Just Because People Are Busy Does Not Mean They Are Productive

Explanation: Without an adequate plan to execute work, a lot of people can be busy doing work that is useless, e.g. developing charts, crunching numbers, designing electrical schematics, writing status reports, developing software code, etc. all of which could be irrelevant if the work is not defined adequately, if work changes or, if the plan changes resulting in rework. I've seen this happen. In fact, I continue to see it. From an earlier example…how about tearing up and rebuilding roads that were not built with a sufficient water control system. As an executive, you need to be acutely sensitive to wasted time equaling wasted money. Project management begins with defining the work, then planning the work, ensuring that the people assigned to do the work are productive.

Figure 7: Wasted Time is Wasted Money

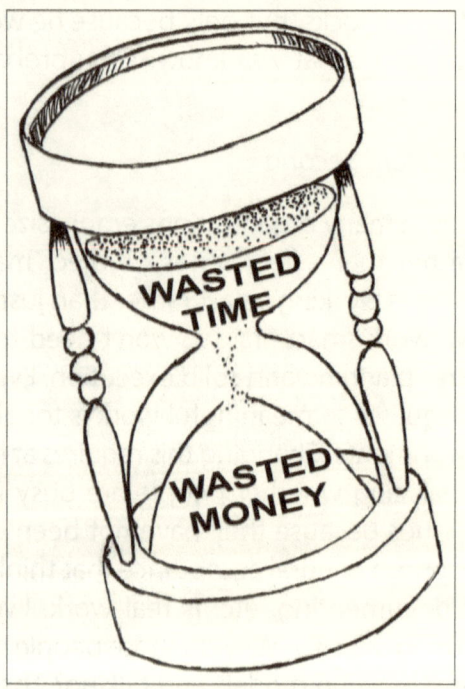

The Immediate Always Delays the Inevitable

Explanation: Even if the work is defined and planned, its status within an organization is always potentially interrupted by other work that has higher priority. This higher priority work is usually in the form of an immediate issue. People are told to stop what they are doing to focus on something else. Understand that an interruption is a delay and any delay is a potential threat to completing projects.

Sometimes the immediate is higher priority. Consequently, strategic, longer-term work such as project work is delayed because of the immediate need. But, here is the main point: The immediate need should be evaluated against the project work to determine actual priority so that consensus is reached and the decision can be documented. Formal project management develops work schedules then maintains metrics based on the schedules. An interruption may be legitimate but should be based on analysis so that the schedule can reflect the interruption. The metrics are not to point fingers, level guilt, or to praise people for completing work early. The metrics are status indicators, i.e. how does reality match earlier expectations and planning? The best way to ensure that your work stays as close to schedule as possible is to assess the relative priority of project work versus the immediate need.

How **Bad** Do You **Want** It, Because that is How **Bad** You'll **Get** It

Explanation: Rushed work is usually substandard, defective and costs more to fix than doing the work the right way in the first place. I was told this slogan years ago by a test coordinator in response to my complaints about sponsors wanting to maintain a schedule by reducing testing. He suggested I ask them, how bad they wanted it because that is how bad they were likely to get it. Of course the initial clause is about emotion while the ultimate clause is about quality. Formal project management includes mechanisms to check emotion so that rationality is maintained and quality can be delivered.

Figure 8: How Bad do you Want it?

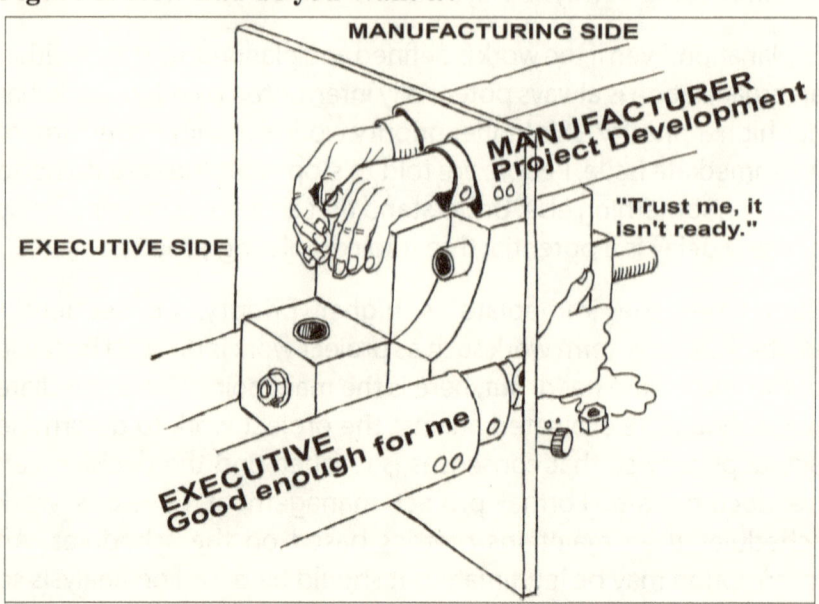

"How BAD do you WANT it because that is how BAD you'll GET it."

Play the Ball, Don't Let the Ball Play You

I use this slogan mostly in reference to methodologies and schedules. Think in terms of baseball, basketball, softball, soccer, tennis, golf – just about any ball sport. Balls are tools and objects to accomplish the goals of the sport. Athletes are more effective when they actively judge the ball's speed, rotation, angle, then adjust to make the ball do what they want: fall into their glove, hit it in the racquet's sweet spot or loft a great pass. They are less effective if they misjudge the ball and get pinned in a corner, shorten their kick or can't reach the ball.

Project methodologies and schedules, like a ball, are tools and objects used to help control and manage work. They should be adjusted based on unique features of the work. Schedules are often adjusted based on changes. One methodology does not fit all types of projects so, adjust. Use these tools but don't let the tools rule you. This does not mean to abandon a methodology phase or let the schedule slip. It does mean that **the project team is in control**, not the tool.

Section 3: Facilitating Success

Characteristics of successful projects need to be complemented with processes, controls and tools that facilitate or help them work. Soft skills presented in Chapter 9 are formal processes that provide efficiency, effectiveness and clarity. The Knowledge Area Topics in Chapter 10 are a short form of the nine knowledge areas from the Project Management Institute. Managing by Metrics in Chapter 11 explains meaningful metrics and how to implement them to manage work. Methodologies in Chapter 12 are structured steps to execute and control projects.

Chapter 9: Soft Skills Rule

The Impact of Facilitation

I was raised in a family with a strong ethic of "a little hard work never hurt anybody." Once we learned to work hard, we learned to work smart. In fact, it was the blood, sweat and tears that prompted me to think of smarter ways to work. Which led me to develop the saying "if we work smart first, we probably won't need to work as hard later."

One experience that remains with me since high school is changing car brakes with my dad one Saturday. These were the old 'shoe' brakes on a 1962 Ford Galaxy. The first brake took us over three hours. We replaced the remaining three brakes in only 45 minutes. After replacing the first brake we were both a bit depressed at the prospect of spending the entire day in the garage. Then, we were elated that replacing the next brake only took 15 minutes. How? Well, we applied all that we learned from replacing the first brake and, learned a bit more with each brake, that we applied to the next brake: These include features of the brake system, how to sequence disassembly and reassembly, which tools worked best in the tight space, etc.

We used the "softer skills" of talking through mistakes to develop a better process; of writing down steps; of where I with smaller hands could reach and hold or my dad with stronger hands to could hold parts secure, etc. As a professional, I have a huge appreciation for what are often called 'soft skills'. Some people will refer to these as 'facilitation' processes. I agree with that characterization. I'm not so concerned with labels as with results. Here is an example from a project I led to conduct scalability testing when an organization with a nation-wide computer network replaced their entire networking system.

The project team was allowed one week at an exclusive and costly vendor test facility. The several-million dollar build out included servers, switches, workstations, routers, software, etc. to replicate network demands for an entire region of the country. Project planning was excellent. The team spent two solid weeks planning all the testing scenarios, specific tests, expected results, measurements, etc. Overall, it was an expensive operation with a tight timeline. We were all stoked and enthusiastic. Midway through the second day, the team encountered an error they could not resolve. I watched them work through possible causes and remediation for about 30 minutes without progress. Knowing how technical people can pursue leads tenaciously while minutes become hours, I realized an intervention was needed to work smarter.

I called to them to get their attention: "Gather all of your data, notes, files…whatever you need to work through the issue. Meet me in one of the break-out rooms across the hall in 5 minutes." We worked through the issue using a structured situation analysis and problem solving system – Kepner-Tregoe. Eliminating false indicators and identifying symptoms that directed us to root causes took about 20 minutes. "We think we know the issues and are ready to implement a fix" was the team's consensus. I was elated they accepted the formal problem solving structure.

They tried the solution, but the problem continued. After about 15 minutes without resolution I called again: "Gather your information and meet me in 5 minutes." After another 20 minutes they had another solution, which actually solved the root cause. We spent a total of about an hour to problem solve something that could have taken all night without a structured approach. Everyone worked together as a team and the resolution process led to clarity and consensus.

Why Soft Skills Rule!

From rules to conduct effective meetings, to structured problem solving and decision making, to ways to influence others, soft skills function as project control factors because they:

- Provide efficiencies.
- Lead to specific definitions, decisions and foundational results that lead to success.
- Guide teams to work smart.
- Allow all parties to contribute without being dominated by strong personalities.
- Provide actual notes, meeting minutes, action items, etc.
- Provide a shared experience and bonding experience for team members, stakeholders, sponsors, etc.

I encourage you, in your role as executive and project sponsor, to both support activities that use soft-skills and to invest in soft skills training for your staff, particularly your project managers. You may encounter resistance from those who believe in working hard and don't see the value soft skills provide to work smart. Some people will say that soft skills are adding needless process into simple activities, e.g. "Hey, it's a simple meeting. Get together, discuss the topics and move ahead." In addition to the examples above you probably have your own examples and, probably have examples where soft skills were not used but should have been. People tend to accept what works so, use various techniques or systems so people can see results.

I have seen decisions that could have taken weeks, resolved in a couple of hours. Issues that seemed very simple lead to fundamental and complex problems which actually saved projects and added value. Not all soft skill processes need to be considered an impediment by adding a lot of structure. In fact, most can be used dynamically, as in the network infrastructure project example. Here is another example of adding structured situation analysis dynamically.

A fellow PM called to inform me that her project team and my project team were vying for the same test environment and that one of our schedules had to flex. The teams could not resolve and both were facing deadlines. To Tina, this was a critical situation. I did not want this to be a conflict among project managers so, I decided to inject structure into the meeting by asking questions like, which area sponsors the projects? Whose budget bought the test environment? Have schedules slipped to create the contention? Because the projects are using the same environment, are they related in some way? Answers revealed both projects had some of the same team members, were sponsored by the same area of the company and were adding different functionality to the same application. Basically, I was guiding them to "define" the problem.

So, rather than debate which team should have first rights, the discussion focused on defining the fundamentals of ownership and value-add to the organization. Within about 15 minutes I told Tina: "This is not a project issue. This is a sponsorship issue regarding prioritization. The sponsoring area needs to determine which project should proceed and deliver first." Tina and the team members agreed, then took the question to the sponsor.

Soft Skills Training

What type of soft skills training are a good fit for your organization? Many exist so, I can't provide an exhaustive list. I recommend consulting your HR and Education & Training staff, both because they should be your subject matter experts, and by showing interest you'll begin to build internal support for soft skills. Reach out to local universities or companies who use various systems. Here are some specific recommendations:

- Overall facilitation training. Great for conducting meetings, structuring working sessions to achieve a result and to navigate through the positional power politics within your organization.
- Situation analysis and problem solving.
    - Ishikawa or fishbone diagramming.
    - Kepner-Tregoe
- Decision making
- Risk and opportunity analysis
- Rules for conducting effective meetings.

As a way to seed your interest and stimulate action, here are two examples of structured processes that can add huge value. Do search for more examples and related soft skill processes.

Ishikawa or Fishbone Diagramming

The Fishbone Diagram or Ishikawa Diagram or Cause and Effect Diagram is named after the inventor Karou Ishikawa as a way to analyze factors that contribute to problems. Collaboration is one great advantage to this process. Although it can be performed solo, teams can use it as ways to brainstorm, identify primary causes and their relative contributions to the problem. Here is an example:

Project Management for Executives

Figure 9: Fishbone diagram examining excessive heat

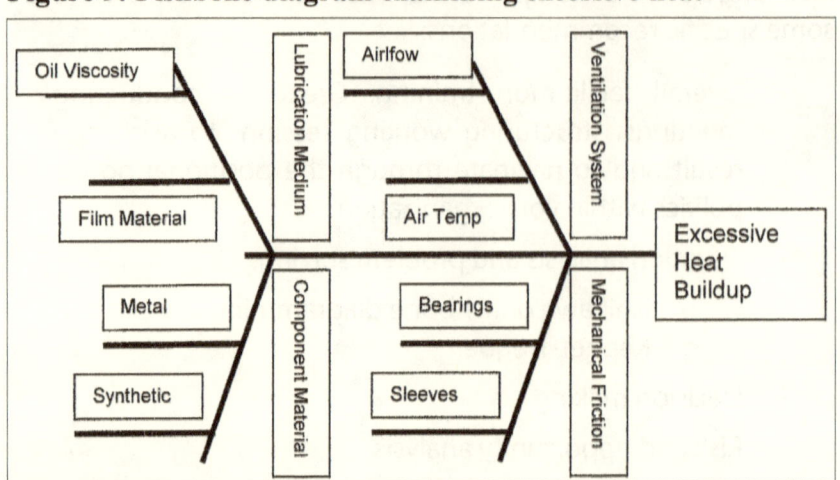

This can be extended and made much more complex with many more bones.

Look at the fairly realistic completed diagram www.siliconefareast.com.

Of course the problem statement, main bones and sub-bones will be unique to your organization. Not only is this a great tool for PMs and team leads but it is also a great reference document for executives like you to appreciate subject matter and situations your staff encounters and must resolve within projects. Why? By now you should be able to answer with your eyes closed: "In order to add value to your organization."

Now, consider using this diagram for planning success. Rather than focusing on a problem, focus on a goal or objective then use the fishbones to list primary success factors, secondary contributors or supporting processes for primary success factors.

Here is an example:

Figure 10: Fishbone diagram for planning success

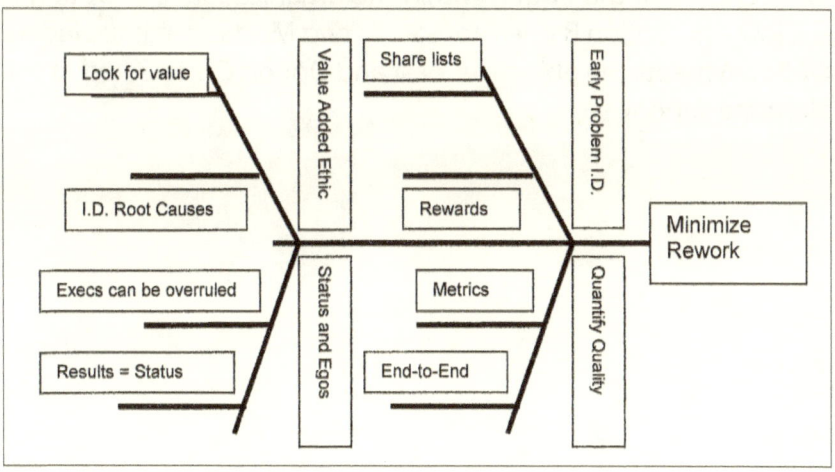

Structured Decision Analysis or Pugh Matrix

I have used the Kepner-Tregoe decision analysis technique successfully in a variety of situations. It is not the only decision analysis technique and I like to say that "There is nothing new under the sun" with it. Within engineering a similar structure is used called the Pugh Matrix. The value of structured decision analysis is the structure and rigor that provide empirical evaluation. However, two key features that all parties involved need to agree are:

- The categories of 'Must Have' and 'Nice to Have'.
- The rankings and values of the options to be decided.

Here is a spreadsheet example. The objective is to select which of Options A, B or C best meet the Objectives, which are classified as Musts or Wants. Must objectives are not ranked or rated because they are essential. Consequently, they are used to exclude options that do not meet them. Want objectives are assigned a value from 1-10 to reflect a relative value. How complete each option satisfies each want is assigned as the Want Rating from 1-10. Scores equal the Want Value X the Want Rating for each Option. Scores are totaled at the bottom. In this example, a perfect score is 350. Option A scores 274, Option B scores 203 while Option C scores 173. Then, a risk and opportunity analysis is performed based on other criteria. Could be how much maintenance the option requires, the relative inexperience of the option within the organization, etc. Option A is a low risk. Option B does not satisfy two Musts so, it is excluded. Option A meets the objectives best and Option C is selected as the alternate vendor.

Table 4: Decision Analysis Matrix Chart

Decision Analysis Matrix Sample										
Decision Objective: Select a new LOB application										
Key: Precise statement. No ambiguity. Objective evaluation and selection										
Options				Option A			Option B		Option C	
Objectives	Must or Want	Want Value (1-10)	Meet Must	Want Rating (1-10)	Want Percent (score X Want value)	Meet Must	Want Rating (1-10)	Want Percent (score X Want value)	Meet Must	Want Rating (1-10)	Want Percent (score X Want value)
Browser based	Must		Yes			No		0	Y		0
Compatible w/existing database	Must		Yes			Y		0	Y		0
Uses same database	Want	5		10	5 * 10 = 50	Y	0	5 * 0 = 0		5	5 * 5 = 25
Individual letter generation	Must		Yes			Y		0	Y		0
Batch/mass letter generation	Must		Yes			Y		0	Y		0
Configurable calculation engine	Want	9		6	9 * 6 = 54		7	9 * 7 = 63		0	9 * 0 = 0
Standard actuarial formulas	Must		Yes			Y		0	Y		0
Create custom actuarial formulas	Want	8		5	8 * 5 = 40		10	8*10 = 80		7	8 * 7 = 56
Integrates w/time reporting system	Want	6		10	6 * 10 = 60		10	6*10 = 60		6	6 * 6 = 36
MS-Windows Server system	Must		Yes			Y		0	Y		0
Security via Active Direcory	Want	7		10	7 * 10 = 70	Y	0	7 * 0 = 0		8	7 * 8 = 56
Security configurable by staff role	Must		Yes			No		0	Y		0
Score		350			274			203			173
Result	Want	Perfect Score			High Score			Eliminate: Two musts not met			Second
Conduct Risk/ Opportunity analysis					Low risk						Moderate risk
Final Decision					Preferred vendor						Alternate vendor

Chapter 10: Knowledge Area Topics

Overview

This Chapter is included only as an orientation to the Project Management Institute (PMI) 9 knowledge areas as key project management subject matter content. Plenty of other books cover these in detail. However, these are not exclusive and other project management organizations use other terms and frameworks. My intent in this handbook is to present their essential features and how they can help complete work and add value. Again, plenty of other books present details to master these topics. This is just not the purpose of this book.

Scope management

Scope management is concerned first with defining the project purpose, goals and deliverables and second, with protecting the scope throughout the project. Scope can be reduced or expanded intentionally and unintentionally during the project. Any changes should be managed via change control processes, which are formally part of integration management. Don't try to rationalize why changes to scope are part of integration. The key point is that changes need to be managed.

Initially scope management processes should define the scope clearly and understandably for the project, sponsors and stakeholders. After this, scope management processes should continually evaluate project requirements and all project work against the defined scope. Consequences of not maintaining scope include exceeding costs, not completing deliverables, completing unwanted deliverables, rework, etc.

Typically four essential documents are created. In your role you should expect your teams to complete these. The value for you is ensuring that the goals and objectives for the project are actually manifested in defined and measurable elements that will be completed.

- A list of all requirements or specifications needed to know, plan and execute the work, produce deliverables, perform testing, etc.
- A requirement traceability matrix (RTM) that traces or matches requirements to deliverables. One deliverable will include multiple requirements. Similarly, multiple tests will probably be performed to demonstrate a deliverable meets requirements.
- A Work Breakdown Structure (WBS) which is a listing of work tasks and activities.
- A WBS dictionary which is a detailed explanation of the work including who is assigned the task, task dependencies, risks and mitigation steps. This document is essential for building a schedule of activities which is often called the project work task schedule.

These documents should be logically consistent because they build upon each other. and must be updated as appropriate throughout the project, i.e. upon approval of change requests. Requiring projects to produce these documents is one of the best steps you can take as an executive and project sponsor. In fact, if stakeholders want to know details about the project beyond the project charter, just refer them to these documents. Using formal project documents is a great way to mature your work culture, i.e. by getting everyone to become familiar with project documents and begin using the common terms, processes, phases, etc.

One classic real example of the value of the WBS Dictionary: I sat on one side of a cube partition from another PM named Jim. I can still hear Jim speaking on the telephone to a project team member. Jim: "I'd like to know why your task is slipping. The schedule shows you should have started it last week." Pause while the team member says something. Jim: "You want me to tell you what the task means? What do you mean you don't know? How the _____ should I know the task. _____ I did not create the task. It's your task...**you created it!!!**"

Now, Jim was right. However, speaking to him about this he admitted that he did not insist the team create a WBS Dictionary so, months after the work was discussed and planned, context has evaporated, memory needs more cues and, no detailed explanation existed to remind the team member. All that exists is a short description on the project work task schedule – which is what Jim is using to see when the task should have begun. Oh well, the task seemed important at the time! Help prevent this on projects by asking your PM for the WBS Dictionary so you can review it.

Time management

Project time has both hour and duration components. All time should be considered estimates. Time is directly related to scope refinement, i.e. as scope is refined from general descriptions to detailed requirements and work tasks, the time needed to complete scope becomes better known and estimates are more precise.

Project management tools categorize time as three types:

- <u>Fixed duration</u> or work that completes in a set number days, weeks or months. Examples include two days to review documents, one week to conduct 'hardening' tests, etc.
- <u>Fixed units</u> or work that is measured and tracked usually as hours, but the unit can be any time unit that is meaningful. Examples include measuring the amount of transactions completed in 50 minutes, translating the cost per hour so that only 40 hours can be devoted to a task, etc.
- <u>Fixed work</u> or work that is defined in units but the duration is dependent on the number of people working. Example: Say a task is set at 80 hours. It can be completed in 4 days if five people work on it or, as long as 4 weeks if only one person can work 50% on it.

and Those Who Want to Influence Executives

Resist the urge to know which tasks are appropriate as fixed duration, fixed units or fixed work. That is the job of PMs, supervisors, foremen, etc. Also, understand that some types of work reach a capacity saturation point. In these situations adding more bodies won't help complete it any faster.

Analysis and knowledge are two key components to time management. Sufficient analysis is required to know the work well enough to estimate work time. Consequently, the project will emphasize analysis and knowledge prior to providing and committing to time estimates. This applies to changing scope. New work or changed work approved through the change request process requires analysis prior to providing work time estimates.

Any slogans come to mind?

I have been squeezed as a PM by clients or other parties who decide they want to move their scheduled dates. On the surface, you may think this is not a big deal. However, any <u>conscious</u> delays to the approved work task schedule impacts dependent tasks, budgeted hours and costs. If delays impact the critical path then the project will not complete on time.

Even if you aren't paying for expensive equipment, you probably have skilled staff who could be doing other things to add value to your organization. Therefore, <u>conscious delays are considered changes that should be submitted, analyzed and approved or rejected</u>. Remember, the work task schedule is an approved document. Even though the time frames are estimates, any baselined work schedule should be considered authoritative. If some party decides they need an extra month to make a decision or they have other priorities, well…all other parties to the project are committed and would suffer the consequences. If consequences don't exist for the delaying parties, then these types of delays are difficult to control.

This highlights the importance of developing the formal Charter and Plan, contracts, SOWS, etc. We'd all like to think that other parties will meet their commitments, but you won't be able to protect your work investment unless you have formal agreements.

People management (HR)

People management includes Human Resource issues so it is sometimes referred to as Human Resource management. People management includes a broad range of people and work performance components including:

- Skills and training required to develop or improve skills.
- Performance assessment and feedback to manage team member performance.
- Work assignment and capacity management. This has two parts:
- First, determining which staff will be assigned to which work. This should be formalized within your organization and controlled by an area or unit that is dedicated to controlling employee work. This will prevent people from being assigned 80 hour weeks.
- Second, matching the amount of work to your organization's capacity to complete the work. Your organization may need to hire more permanent staff, temporary staff for limited assignments, etc.

I've seen a lot of organizations who don't understand why their staff seems to bounce from job to job or work assignment to work assignment. Formalizing how staff is assigned can really help staff focus on fewer tasks, which usually results in higher quality and faster completion.

One of your jobs as an executive and project sponsor is to help resolve work assignment conflicts. If the same people are needed by separate projects during the same duration, well something has to give. How do you decide? This is the question of work priority. Each project manager thinks their project is top priority. Perhaps each immediate sponsor thinks the same. As an executive sponsor, you may be asked to resolve the impasse.

Similarly, I've seen organizations not control the amount of work, which soon overwhelms their staff's capacity to do it. Commitment dates are missed and quality suffers, often resulting in more fixes or repairs over longer time frames.

Quality management

Quality management is concerned with satisfying specifications. How hard should the concrete be? How many seconds are needed to return a search query through a web application? How cold should an electrical component be able to function? How many cars can move through downtown for 1 mile if the lights are synchronized?

Because quality can be a broad term and also have very narrow application, the quality expected within deliverables, communications, etc. should be specified <u>where it is important</u>. Two keys are:

- Avoid a sliding scale of quality that creates uncertainty and rework.
- Provide sufficient specifications to measure when deliverables can be approved and when work is complete.

Quality specification is a factor in scope management. If quality is not defined sufficiently then scope may not be met. However, if quality is defined to an extreme level then scope may be exceeded which may also require longer time and increased cost. Project teams will perform quality definition and specification analysis, then will work to meet these quality specifications.

My wife's uncle worked for Hewlett-Packard and was very successful acquiring government contracts. I was complaining about $50 bolts and $1,000 toilet seats once. Alan politely explained to me that quality specifications often require special manufacturing, materials and processes, all which increase costs well above mass produced parts that look almost identical. A fifty-cent bolt will be fine for a homeowner replacing a bolt on a garage door frame. But, applications that require high load strength and withstanding high vibration frequencies may need the $50 bolt. I'm not saying to reduce quality, just to manage it appropriately.

Sample team member quality activities should include:

- Development steps for business and technical components.
- Peer reviews of requirement statements, designs, solution options, etc.
- Updates to documentation like IT artifacts, blue prints and landscape plans.

Communications management

Communications spans a broad range from formal reports to informal agreements, i.e. results of conference calls and emails. Projects should focus on formal communications and communications that result in decisions that affect project scope, time and cost.

Examples: The project will manage formal meetings, conference calls, reports and report formats. The project will also manage decisions reached by team members that are ad-hoc in nature. Examples include: Verifying via a project document resolution of some issue. Documenting an understanding or decision reached via a conference call or series of emails. In essence, if the nature of communications is important, then it must be documented. <u>If not documented then decisions or agreements will not be binding</u>. Your role as an executive to enforce formalizing ad-hoc communications can really help stabilize projects and reduce the potential for negative impacts.

Require projects to develop a consistent and repeatable status meeting and status report format. Even better, encourage adopting a consistent format within your organization.

Documentation has a time Impact. Team members should be expected to build time into their work task estimates to account for communications. If a four-hour meeting requires an additional hour for documentation, then team members should build five hours into their work estimate. As an executive, don't underestimate the value of this. Here are some other examples of formal communications.

- Communication components and formats
- Reports
- Metrics
- Controlling documents
- Approvals
- Memos of Understanding

Risk management

Risk management focuses on unexpected events that can impact projects. Risks should be evaluated dynamically and continuously throughout the project. Team members are empowered and expected to identify risks and to document risks. Here is another place executives can help. Often, team members are reluctant to raise risks, let alone document them. Why? Fear of being viewed as negative or holding up work for something that may not happen. You should let everyone know that they are expected to state and document risks.

Each risk should have one or more mitigation strategies, also known as actions to either reduce the likelihood the risk will occur or, to reduce the severity of the risk. Some mitigation strategies will be integrated into the project as proactive and protective work tasks. Risks should also be evaluated within the WBS dictionary for each work activity.

A variety of risk tools exist to help evaluate both the probability a risk will actually occur and the impact if the risk does occur. Set the expectation with your project managers that they use a risk evaluation matrix and report risks on status reports and during status meetings.

Cost management

Cost management focuses on the budget as money, hours for

work, or both. Think of this knowledge area as accounting for projects. Essential components include how a project is funded, the amount of funding, funding for specific components, i.e. hardware, software and, staff and funding changes during the project. Cost management interacts with scope, quality and time, particularly when costs are fixed and the project must deliver what it can without exceeding cost.

Cost management will include regular metrics within regular reports, as agreed by sponsors and the project management. Think about the type of metrics you'd like to see. If you needed to provide cost accounting for project work, what type of metrics would you need?

Project cost accounting metrics are a great sign of project planning and executive commitment to project management. A lack of project metrics indicates the project was not planned so that metrics could be measured, collected, analyzed and used to evaluate project health. Again, many organizations think taking the time to plan delays real work. Just get working so we can see progress!!! Remember the slogan "just because people are busy does not mean they are productive." If your organization measures hours charged against a work task code or the number of hours a forklift moves material, then your organization is not measuring the right metrics. The right metrics are evaluated against scheduled work that has been sequenced and estimated. Remember that determining these items are real work. The result will be meaningful metrics.

Requests to exceed the project budget must be submitted to project sponsors via the change request process, with an adequate amount of time for review, decision and subsequent action; including reducing project scope to accommodate budget. This is when your role as an executive and project sponsor need to determine if continuing the project is worth the investment. But… how will you know if you don't have cost accounting that allow you to evaluate metrics such as:

- Cost of the project to date.

- Cost spending rate based on the progress to date.
- Comparison of the projected cost from the current point to project end. This needs to be compared to the original budgeted amount from the current point to project end.
- What is the estimated cost to complete the highest priority deliverables but delay the nice-to-have deliverables until the next budget cycle.

You get the point.

Many metrics exist for cost to evaluate the rate a project is using its budget. How much continuing the work through completion will cost from a certain point. Metrics are important because they allow you as an executive sponsor – managing project work relative to an organizations total portfolio – to determine whether to continue a project or stop it. Insist that projects take the time to plan the work and plan to measure the work. Then, you can insist on receiving meaningful metrics.

Procurement management

Procurement management is concerned with purchasing products and services. Consequently, contracts, SOWs, confidentiality and non-disclosure agreements are essential components. Projects should not develop their own procurement documents. Rather enterprise or company-approved agreements should be used and either the purchasing area and/or legal department should be involved in developing and approving procurement documents. Be sure to inform PMs of all organization requirements for procurement-type documents, because agreements may not be valid if not approved through proper channels nor approved by management with signature authority.

Typical contract types include fixed price, time and materials, cost-plus (fixed initially followed by sliding scale or other). Purchasing and legal can provide advice on the most beneficial contract type and which provides the best opportunity for managing risks.

Procurement management interacts with cost management and risk management.

Some general rules for procurement documents that should be shared with project managers:

- Protect your organization through terms, conditions, what-if scenarios, etc.
- Include incentives for contractors or vendors to perform with quality and to stay on schedule. I really don't like using holdbacks but not paying 10% to 15% until all the work is complete at a satisfactory level does provide a strong incentive.
- Include penalties when contractors or vendors do not perform with quality or stay on schedule. Even though delayed payments are often sufficient incentives for contractors to stay on schedule, performing with quality may require small-scale demonstrations or measurement of pre-work to evaluate that quality is at a sufficient level.
- Require schedules, bonds, qualifications of staff, policies and procedures, etc.

Integration management

Integration management is concerned with coordinating all aspects of the project. Notice statements in the other eight project management components about components interactions. Well, integration management is about managing how they do interact and how to pull together processes, tools, schedules, etc. cohesively and seamlessly.

Change Management is an essential part of integration management. Although it has been addressed several times, there is one last set of information to present.

All change requests will be reviewed via a four-stage process:

Stage 1: Requests. Changes are documented and submitted for

consideration. Usually a formal and standardized change control document is used. Note that requests may or may not be approved. Executives and sponsors may not need to be on the change board. However, some changes may need to be escalated for awareness and approval.

Stage 2: Analysis. The request is analyzed to determine impact to work effort and feasibility within project constraints. This is important to ensure the change can be made and to ensure the change is consistent with project scope. This is another point during which items may be escalated to you as sponsor. If the change deviates from scope then project teams should reject. At this point, those wanting the change may want to escalate and you are on the escalation path. Just because the project has time, people and budget to make the change does not mean it is in scope and should be approved. Your job is to protect the project and this includes saying "No".

Also, the number of parties involved may require multiple levels of approval so, there could be Stage 2-A, Stage 2-B, Stage 2-N.

Stage 3: Approval and communication. I include these together because changes should be communicated to the project team, sponsors and stakeholders. If the request and analysis stages are conducted properly and with all appropriate parties, no one should be surprised.

Stage 4: Reset the baseline. This includes updating all appropriate controlling documents. **Note: The change management process is not complete until Stage 4 is complete.**

Management and escalation paths

Decisions need to be made with every project, which create opportunities to approve and deny, say 'yes' or 'no' and for those who disagree with a decision to request escalation to a superior manager. You need to support your project manager's ability to manage a project, which includes making tough decisions. However, situations do exist where decisions do need to be escalated. Be sure that if you counter the decision of your project manager that you do so in a way that does not diminish the PMs status and authority for future decisions that need to stop at the PM.

Many organizations fail to implement and document escalation paths. Sometimes they don't want to butt heads with clients. Sometimes they feel they don't have the positional power to escalate items or say 'no' if a decision is escalated. While I understand those circumstances and their foundational emotion…and it is emotion…not implementing escalation paths can lead to even worse situations. <u>Your role as an executive within your organization includes this function for projects</u>.

Chapter 11: Managing by Metrics

Value of Metrics to Success

In some respects this chapter is longer than I intended for this type of book. On the other hand, metrics are soooo important that I wanted to err on the side of more information than less. Take your time reading this. Use your own data to work through formulas. I think you'll understand the value.

Metrics are extremely important in an organization. Probably second only to the organization's core business itself. Metrics exist at a financial level to demonstrate revenue, expenses and profits. They exist at asset levels for inventory, product flow, order triggers and shelf-life. Metrics also exist at the level organizations need to measure aspects like:

- How effectively staff work to complete deliverables, whether a print run or installing an assembly system.
- Whether expenses remain high for several months following a project because defects are found and must be corrected.
- Reinforcing the value of processes to guide work.

You can probably think of many metrics you wish your organization generated that could add value. An entire set of metrics exists within project management. Here are some samples:

- <u>Schedule progress</u>, i.e. will the project finish on time or how far ahead of schedule is it?
- <u>Cost comparison</u> to an overall budget, per contracts, per deliverables, etc.
- <u>Quality measurements</u> that indicate whether what is being build meets specifications.
- <u>Hours</u> spent to produce each deliverable.

Think of these project metrics as analogous to operational and finance metrics. Your organization should take time to determine which project metrics you want to capture. Then you can plan out the details for capturing, reporting and making decisions based on the metrics. Honestly, implementing metrics is one of the most valuable accomplishments for your organization. Benefits will include that your organization can be managed by metrics, by facts, rather than by hunches and anecdotes. Also, project managers actually relish the opportunity to plan projects to the extent that metrics can be calculated, reported and used to manage work. Partner with your project managers to do this and watch your return on investment climb.

Standard Project Management Metrics

Now, I want to present standard project management metrics, then some ideas for custom metrics. These standard project management metrics require several foundational factors.

<u>First, sufficient planning</u> so that work is defined, estimated, sequenced and scheduled by the people who will do the work. If project tasks are too high-level or created by someone other than those who know the work, then a set of 8 tasks that are scheduled to complete in one week can easily grow to 18 tasks that take a month and involve five additional people. Trained project managers can facilitate effective planning, so give them the time to do plan.

<u>Second, a commitment to record keeping</u> so that data is current and accurate throughout the project. This is a simple "gold in, gold out" versus "garbage in, garbage out" principle. Metric accuracy and meaningfulness is dependent on the accuracy of the data. Options exist for producing fairly accurate data even if your organization does not have automated or precise methods to record time. For example a spreadsheet can be used to list work tasks to the nearest 15 minutes. This is an arbitrary interval so, you can make this 5 minutes or a tenth of an hour if you want. Making record keeping straightforward, easy and meaningful for employees is the best way to develop their commitment to it. The more difficult or

confusing the less likely employees will use it and the less accurate the data. Also, don't worry too much about rounding to the nearest N minutes. Based on the law of averages and the principle of a generalized distribution around an average, this rounding should work pretty well.

Third, a reasonable amount of time and effort to produce the metrics. You may be able to automate producing some metrics while others will need to be calculated either by hand or via formulas entered in a spreadsheet.

Again, set expectations within your organization for these contributing factors. As your organization matures its work environment, these contributing factors will become more ingrained and part of managing work.

The Fence Example

Let's use a simple example of fencing a lot. Installing the fence around all four sides is estimated to cost $12,000 and take 12 days, or 3 days per side. This equates to a budgeted value of $1,000 per day and $3,000 per side. If each post and section of fence represented $100 then each side would have 30 posts. Here is a visual:

Project Management for Executives

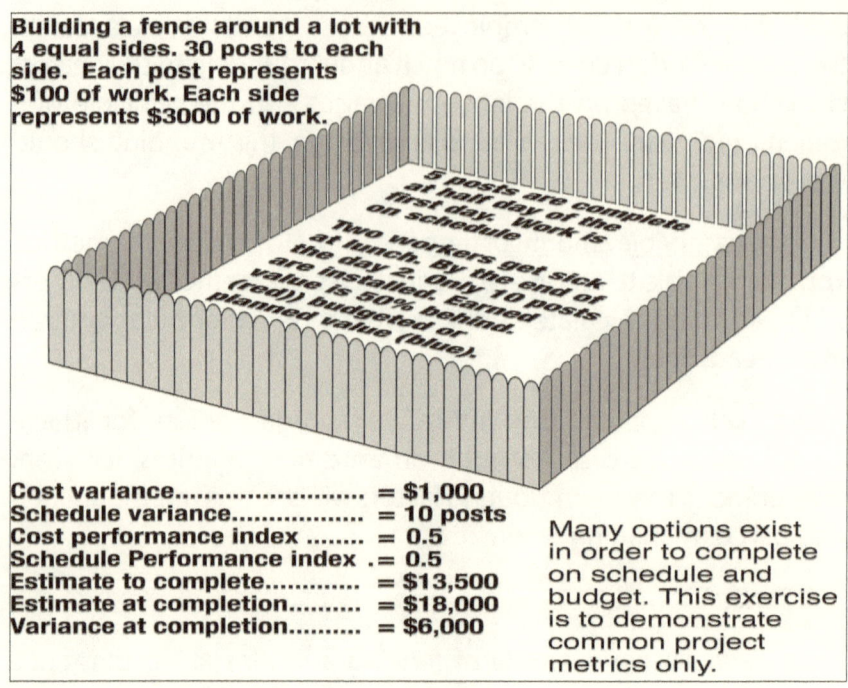

Figure 11: Metrics Fence Example

This example is a simple but great visual. Draw your own then play with various scenarios to progress work on each side until the job is complete. Use the metrics below for calculations and status for your scenarios. As you do, compare to how work in your organization progresses and how to identify metrics along the way.

Budget at Completion: The budget to complete the work or $12,000.

Earned Value: This is the amount of value that has been 'earned' to a certain point based upon the estimated budget. This statistic involves both cost and schedule. A daily rate of 10 posts would equal 100% of earned value. Say that at half-day or lunch of the first day 5 posts were installed and represents $500 of value. The project is on schedule.

However, two workers get sick from lunch and miss the rest of that day and the second day, so that only 10 posts are installed at end of day 2. Earned value = 10 X $100 = $1,000, instead of the $2,000 value expected by end of the second day.

Cost Variance

Cost variance is the difference between budgeted value and earned value. Earned value at day 2 equals $1,000 - $2,000 = -$1,000. The project has spent $1,000 more to this point than budgeted. Cost variance can be calculated at various points through the project. If project work can be spread across all work fairly evenly – sometimes a big assumption – then both earned value and cost variance can be determined at a very discrete level, e.g. per day.

Schedule Variance

Schedule variance is the difference between the expected or budgeted and actual duration; often in terms of hours, days or weeks. In this case the budgeted schedule would be completing 20 posts at the end of day two. The variance is calculated as actual (10) minus budgeted (20) = -10. The project work is behind by 10 posts.

Cost Performance Index

Cost performance index is the percent of budgeted cost to a certain point compared with the actual or earned cost. The formula is actual / earned. In this case the cost performance index = $1,000 / $2,000 = 0.5 or 50%. A perfect index would be 1.0 or 100%.

Extending this out to day three, the budgeted or expected budget is $3,000 which can be multiplied by the CPI of 0.5 = $1,500 at the end of day three.

Schedule Performance Index

Schedule performance index is the percent of the scheduled work that has been completed. In this case 10 posts are completed compared to 20 posts so, the schedule performance index = 10 / 20 = 0.5 or 50%. A perfect index would be 1.0 or 100%.

Why are cost performance index and schedule performance index important? Because these metrics can help indicate the status of the work in order to manage the remainder of the work by doing things like, extending the completion date, adding more workers to catch up, getting more power equipment to dig holes and install posts faster or asking for more money. Managing by metrics is a big deal, but can't be done if the metrics are not calculated.

Estimate to Complete

Estimate to complete is calculated by extrapolating the earned value for the remainder of the project by using the cost and schedule performance indexes. If the two sick workers do not return or are not replaced, the rate of work would reasonably continue to be half of the budgeted work every two days. The project is scheduled to cost the original $3,000 by end of day three, plus $1,500 additional dollars for a total of $4,500 for side one.

The remaining three sides would also cost an additional $1,500. Calculated:

Budgeted cost for sides 2, 3, 4 = $3,000 X 3 = $9,000.

Estimated cost increase for each side = $1,500 X 3 = $4,500

Estimate to complete the remainder of the work = $9,000 + $4,500 = $13,500

Estimate at Completion

The estimate at completion is the total of the estimate to complete plus the costs to that point, also called sunk costs. In this example: $13,500 + $4,500 = $18,000.

Variance at Completion

The variance at completion is the budget at completion minus the estimate at completion. In this example $12,000 - $18,000 = -$6,000.

Cumulative Cost Performance Index

Cumulative cost performance index is the cumulate percent at successive stages of work. This could indicate trends that are arithmetic or geometric. Often graphs are terrific tools to present these types of relationships. For our simple example, the cumulative cost performance index is equal to the standard cost performance index.

Return on Investment

Return on investment is based on the value of the completed work over some duration. Let's say the fenced lot is to store securely recreational vehicles like boats or motor homes or, to provide a large dog park for exercise. The return on investment would equal the sunk costs of the deliverable or project at completion divided by the income. Recovering the sunk costs may take 12 months with a steady profit after month 12.

Internal Rate of Return

Internal rate of return is the financial return compared to other internal costs, i.e. loans used to construct the fence. If the revenue from renting the lot = $1,000 per month and your organization is paying $800 per month for the loan then the internal rate of return is $200 per month.

Make Metrics Practical and Meaningful

The meaningfulness of these metrics really depends on each project. As much as I promote metrics, I also say that they are tools to help people manage things; in this case manage work. So again apply the slogan – play the ball, don't' let the ball play you. You are in control. Use the metrics but don't let the metrics lead you into forcing their use where not really appropriate.

All metrics are great for managing large and complex projects. However, some are not really worth the overhead for small and uncomplicated projects. Examples:

- A project to conduct an employee opinion survey during two months within a small organization, say less than 200 people, could benefit from some straightforward metrics like cost variance and schedule variance. However the project would probably be near completion before metrics like cost performance index and schedule performance index would help manage the work. That said, all metrics could be used after a project completes to evaluate how well the work was planned and executed. Evaluation like this can be used to plan and execute the next project more effectively.

- A project to design and install a fire-suppression system into an existing five-story building could benefit from all of these metrics.
- <u>It would take long enough</u> that all scheduling metrics would provide value.
- <u>It would involve purchasing enough materials</u> to justify using budget and cost metrics.
- <u>It has sufficient improvement value</u> to calculate return on investment and internal rate of return metrics.

So, the point is that one size does not fit all. Use the metrics that fit the nature of the project. But, DO USE METRICS. Hold project managers and sponsors accountable for defining, gathering the data and, calculating the metrics that will be used during the project. Again, everyone in the organization should understand the value of metrics and do their part to provide the data needed to generate the metrics. This may be by recording work hours as accurately as possible each week or by saying "those facilitating processes really helped the last project move faster. Let's use them again to see if we can stay ahead of the planned schedule." Also, if managers in your organization reviewed metrics with employees the value of the metrics would be reinforced, as well as the expectation that employees do their part.

Another huge value of using metrics is reinforcement of the three key factors listed earlier:

- Sufficient planning
- Commitment to record keeping
- A reasonable amount of time and effort

Your organization will see that these really are key factors and begin building them into their daily work, project planning and estimating. People will also breath a sigh of relief knowing these actually help add value to their organization.

Additional business and operational metrics can be incorporated into a project management work environment. For example, Return on Investment (ROI), Internal Rate of Return (IRR), sunk costs, etc. These are valuable for evaluating projects within a total investment of the organization or Portfolio Management.

Custom Metrics

Two colleagues from different organizations are in a cross-company mentoring program, and are comparing features of how work is accomplished in their respective organizations.

Dean: "We were able to start 10 separate work efforts last year (let's call these projects) that we estimate represent 40,000 hours of work."

Teresa: "Wow, you started 10! How long did the initiation time take? We've found a strong relationship between prep time and completing within 10% of the estimated duration." We decided that because completion represented gaining value from money we already spent, i.e. our sunk costs, that we benefit more from starting only the work that is planned sufficiently and that we know we have the staff capacity to complete.

Dean: "Don't know the initiation time. We pull in several executives and unit managers for a couple of one-hour discussions, then they use some templates we developed to describe and estimate the work. I'm impressed by you guys completing work within 10% of estimates. Ours is 50% or greater. How many projects do you start each year?"

Teresa: "We increased each of the past three years. First year was low, three projects, because we learned what planning well meant the previous two years. We only start as many projects as the teams can plan adequately. Planning to a level of high quality has improved each year. Second year we initiated 5 projects and still had capacity for more. This year we started 6 projects and that is about our capacity. We found that employees in the trenches actually provided the best planning information, rather than executives and unit managers alone."

Now you are pretty smart, so you can see where this scenario is headed. Custom metrics can help your organization evaluate some critical success factors of completing work to add value. Just as in this scenario, your organization can compare up-front planning with end results. This type of metric can help guide your operations by showing the value of planning. However, not capturing this type of information results in relying on anecdotes and hunches.

I recommend using a two stage approach to determining custom metrics:

1. Rate the metrics as those essential or highest priority to lowest priority.

2. Incorporate them into small projects first so you gain experience; then you can institutionalize them within your organization.

Chapter 12: Methodologies

Overview

Methodologies add specific steps, sequences and rules of operation for completing project work. All methodologies are approaches of how to complete work. The nine knowledge areas are used within and complement a methodology. I'll make a bold statement: All methodologies will work, but not all are most appropriate, depending on the type of work, industry and subject matter area. Presenting some methodologies is the best way to clarify this. However, this is not exhaustive and other books can provide much more detail. I recommend not getting hung up debating or deciding on a methodology. Just about all can be made to work… but **use a methodology!**

Waterfall Methodology

Think of work progressing through a series of waterfalls. Work tasks are grouped as high-level categories that progress sequentially. Work in a downstream category does not begin until all the tasks are complete in an upstream category. Consider a bridge maintenance project. Sequential categories could be:

1. Planning
    a. Evaluate structural integrity.
    b. Identify maintenance activities.
    c. Monitor traffic activity.
2. Prepare for bridge work
    a. Order parts and equipment.
    b. Determine traffic control operations.
    c. Prepare bridge sections for maintenance.
3. Development
    a. Structural integrity maintenance work.

b. Surface preparation and painting.
4. Clean up
   a. Remove equipment.
   b. Remove traffic control

Figure 12: Waterfall methodology

Category 1 Planning tasks would need to complete prior to beginning Category 2 Prepare for bridge work tasks and the same for Development and Clean-up tasks.

While this provides sequential controls, the level of control in a waterfall methodology is not always practical. For example, your project team may need to place orders for some parts and equipment months ahead of completing all Planning tasks. Otherwise, the time needed to receive the parts and equipment could really delay work. Similarly, a traffic control plan could be developed based on recent historical information from a county or city without needing to wait for all Planning tasks to complete.

Waterfall is very effective for many smaller, limited-scope projects because it is logical and not much time is wasted. However, even a waterfall methodology can have several phases running concurrently. The key is accounting for any dependencies across work phases, particularly for projects with moderate to high complexity.

Agile Methodology

An Agile methodology provides flexibility to perform work tasks based upon dependencies for each task. Also, the elephant of a larger project is eaten one segment or deliverable at a time within short duration iterations. Each work segment or deliverable is defined, planned, built, tested, delivered, then the team moves to the next work segment. Using the bridge maintenance project example again, task 3 in Planning: <u>monitoring traffic activity</u> is not dependent on <u>evaluating structural integrity</u> so, it can actually begin at the same time or earlier. Similarly, from the outset your team knows sandblasting equipment is needed to prepare metal for painting. Because sand is readily available, an adequate amount of sand can be ordered immediately with more sand delivered as needed.

Agile is a type of methodology developed and used primarily for software development, but many organizations have modified it to fit the nature of their work. Many great resources on the internet describe Agile in detail. Purists may wince at my description, but information from purists are readily available for when you need that level of detail.

Sprints

Sprints are an implemetation approach often used with Agile that focuses on very short deliverable phases. Key features are:

- Deciding on the length of a deliverable iteration, or Sprint. Thirty-days, 60 days, 90 days, etc.
- Deciding on sequencing deliverables, i.e. which comes first, second, third, etc.

- Small teams that work together to plan, build, test, deliver, etc.
- Low overhead because the work is divided into short deliverables.

Many lengthy descriptions are available, complete with rules, steps, tracking, team roles and authority, etc. I really like Agile for three main reasons:

1. It focuses on discrete deliverables so, success is evident, measurable and self-reinforcing.

2. Focusing on discrete deliverables causes everyone to think in project management terms of defining the work, planning the work, protecting the work from risks and changes and, delivering the work within the time frame.

3. Changes to an existing deliverable can be incorporated during subsequent work phases. This is a great change control mechanism.

Other benefits include:

- Sprints can actually fit repeating work cycles also referred as 'releases'. Let's say your organization implements new features, products, prototypes, etc. on a monthly or bi-monthly cycle. Sprints are ideal for adding the rigor and structure needed to manage work within these cycles, reducing implementation slippage.
- Agile complements work capacity management. Staff can cycle through a Sprint deliverable then move on to maintenance work. Each phase is short enough that planning the next phase against the available staff and other work is very manageable.

Many other methodologies exist and are specific to industries or more appropriate for manufacturing or research and development operations. Some companies use multiple methodologies. Other systems exist to manage work and are proven quite successful. Six-

Sigma and Lean are two others. Each have their processes, roles, authority, metrics, etc.

So, from these fairly popular methodologies you can see that any methodology can work. The best methodology for your organization is up to you. One company I worked for used a custom methodology developed by a consulting company. Even though it was well documented and training was provided, it contained so many options, so many steps and so much repetition that its complexity created paralysis by analysis. I encourage you to find an existing, industry accepted methodology and add as few unique features as possible.

Why use an industry accepted methodology? First, your organization will be able to work much easier with other organizations, such as vendors, clients and consultants, because they will also be familiar with industry standard methodologies. Second, plenty of 'how to use' information and training is available. Paying for this type of training is far less costly than developing your own methodology or using one that is not industry accepted.

SCRUM is a team approach often used with Agile

- Small teams (5 – 9 people) with defined roles:
- SCRUM Master, facilitates SCRUM sessions
- Product Owner (ownes the prioritized list or "backlog" of requirements
- Teams meet daily for 10 – 15 minutes
- SCRUM agenda:
    - Each person anwers 3 questions:
        - What did you accomplish yesterday
        - What are your plans for today
        - What issues are inyour way, or are you about to impact anyone else?

Closing Comments

I have mentioned that purists may disagree with how some content is presented. Also, some content does not align precisely with recognized project management organizations. Doing so is recognition that great processes and principles can and should be adapted, i.e. "morphed" to fit a particular work environment or industry. The content of this book does support applying sound principles so that your work can succeed to provide value to your organization.

Also each work environment or context has unique technical and work management features that require morphing of principles and processes. For example, pension management organizations are vastly different from road construction companies. A technology research division of a company has quite different employees and project delivery demands than a custom steel fabrication company or agribusiness.

I argue that the strength of project management processes and principles are their ability to be morphed yet still provide value. So, while my eight steps of work and 20 characteristics of successful project work environments do not match exactly what others advocate, I expect that you will morph what I present to fit your needs.

An important goal of this book is to present ways to influence your organization. Ways to get employees to understand how they can work smarter. A question I asked all reviewers is: "Do you think the active exercises and slogans are appropriate and could be used in your work environment?" Most reviewers thought they could with this caveat: "We've never done this sort of thing before, so I don't know how it would be received."

Project Management for Executives

DO NOT LET THIS DETER OR DELAY YOU! USE THESE AND OTHER ACTIVE EXERCISES!

Organizational change only happens if the change reaches a critical mass that can then self generate. Using this book among a small group of people is a good start but will not result in reaching that critical mass. That also requires a plan to get your people thinking and talking. Get them to the point where they are willing to try new things and discussing what worked, what did not, and challenge each other to think in terms of adding value. Get your executives to the point where they are willing to give up control to a project team, feeling confident that the framework for success is so strong that the project team can succeed with oversight and light touches, but not intrusion.

Success breeds success. Go for it.

Appendices

HELPFUL TOOLS

Top 20 Success Factor Table

Success Factors	To What Extent are these Success Factors In Place? Low to None = 0 or 1 Somewhat or sometimes = 2 Mostly but could be better = 3 Yes or appropriately = 4
1. Executive sponsor engaged.	0
2. Immediate sponsor engaged.	1
3. Project manager engaged.	2
4. Client / customer engaged.	3
5. Project control framework accepted and used.	4
6. Formal charter stating the project mission..	3
7. Deliverables are defined.	2
8. Stakeholders established	1
9. Management plans developed and authorized.	2
10. Core team established and allocated appropriately.	3
11. Project team has autonomy.	4
12. Reporting and metrics established.	3
13. Ethic of adding value to the organization in place (quality, changes, risks, issues).	2
14. Escalation paths defined & encouraged.	1
15. Progression of: scope to req. to work to schedule followed.	1
16. Contracts, SOWS, NDAs, etc. in place.	2
17. Change management established	3
18. Risk management established	4
19. Methodology to guide work in place.	3
20. Soft Skills / facilitation processes in place.	2
Score	45 = Questionable
Success Rating	Score = 0 – 29 Bad (Are you prepared for the cost of chaos?) Score = 30 – 49 Questionable (What is the cost of getting to 'Better'?) Score = 50 – 69 Better (Issues are probably manageable.) Score 70 – 80 Best (Architected for success.)

Project Management for Executives

Facilitation Tools

Decision Matrix

Decision Analysis Matrix Sample
Decision Objective: Select a new LOB application
Key: Precise statement. No ambiguity. Objective evaluation and selection

Options	Must or Want		Option A			Option B			Option C		
Objectives	Must	Want Value (1-10)	Meet Must	Want Rating (1-10)	Want Percent (score X Want value)	Meet Must	Want Rating (1-10)	Want Percent (score X Want value)	Meet Must	Want Rating (1-10)	Want Percent (score X Want value)
Browser based	Must		Yes			No		0	Y		0
Compatible w/existing database	Must		Yes		0	Y		0	Y		0
Uses same database	Want	5		10	5 * 10 = 50		0	5 * 0 = 0		5	5 * 5 = 25
Individual letter generation	Must		Yes		0	Y		0	Y		0
Batch/mass letter generation	Must		Yes		0	Y		0	Y		0
Configurable calculation engine	Want	9		6	9 * 6 = 54		7	9 * 7 = 63		0	9 * 0 = 0
Standard actuarial formulas	Must		Yes		0	Y		0	Y		0
Create custom actuarial formulas	Want	8		5	8 * 5 = 40		10	8*10 = 80		7	8 * 7 = 56
Integrates w/time reporting system	Want	6		10	6 * 10 = 60		10	6 *10 = 60		6	6 * 6 = 36
MS-Windows Server system	Must		Yes		0	Y		0	Y		0
Security via Active Directory	Want	7		10	7 * 10 = 70	Y		7 * 0 = 0		8	7 * 8 = 56
Security configurable by staff role	Must		Yes		0	No		0	Y		0
Score		350			274			203			173
Result	Want	Perfect Score			High Score			Eliminate: Two musts not met			Second
Conduct Risk/ Opportunity analysis					Low risk						Moderate risk
Final Decision					Preferred vendor						Alternate vendor

98

Iron Triangle and Three-Legged Stool Graphic

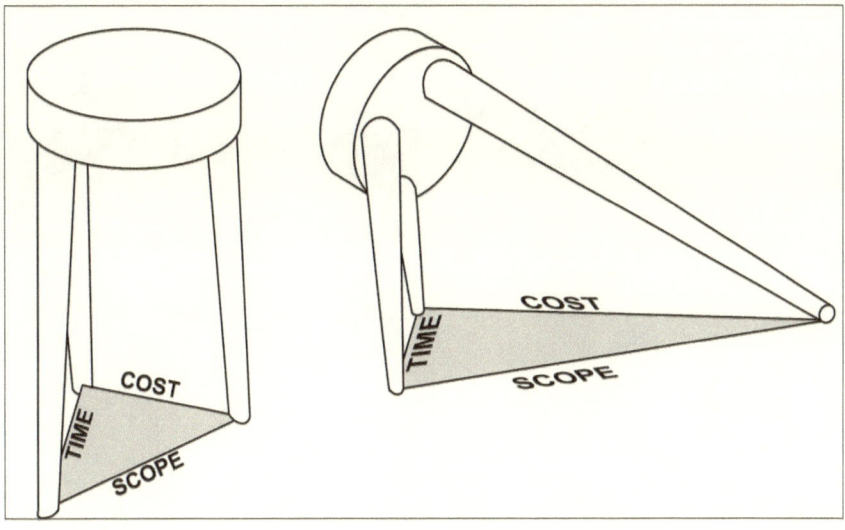

Index

A

As an Executive
   You are in an ideal position 1, 2
   You need to emphasize and insist that processes be developed 4
   You need to ensure the work is defined 2
   You need to make prioritization decisions 3
   You need to support the basic principle that planning is real work 4
   You need to support the project entity 5
   You need to support the time and effort to figure things out up front 3

B

Book Sections ix, 1, 51

C

Change Management 46, 76
   Stage 1 Request 76
   Stage 2 Analysis 77
   Stage 3 Approval and Communication 77
   Stage 4 Reset the Baseline 77
Companion book x, 11, 47
Companion Book
   A Fresh Look at Improving Your Work Environment Using Project Management Principles x
   Creating a Value Added Work Environment Using Project Management Principles 47

D

Defining Success
   Success Factor Explanations 38
Demystifying
   Demystifying Project Management 7
Demystifying Project Management 7

E

Exercises and Activities

Project Management for Executives

    Develop a definition of success 20, 32
    Development exercise 16, 46, 51
    Prioritization exercise 10

F

Facilitating Processes
    Cause and Effect (Fishbone or Ishikawa) 61
    Decision Analysis 63
    Decision Analysis Matrix 65
    Facilitation 57, 98
    Fishbone Diagram (Ishikawa) 61
    Ishikawa Diagram 61
    Pugh Matrix 63
    Soft Skills 44, 57, 59, 60
Facilitation Tools
    Decision Matrix 98

I

Ideal Work Steps 15
    Acceptance of the project charter as the official mission 13
    Determine what the organization wants the project to deliver 13
    End the project 15
    Formally accept project deliverables 15
    Run the project 14

K

Knowledge Areas 43, 66

M

Managing by Metrics
    Managing by Metrics 42, 57, 79
Methodologies
    Agile Methodology 92
    Scrum 92, 93
    use a methodology 90
    Waterfall Methodology 90

O

Operational Definition of Project
    Definite end 19
    Definite start 19
    Projects as an entity 18
    Specific set of work 18

P

Perspectives 6, 16, 51
  Demystifying 7
  Doing the same things repeatedly knowing you'll have problems 16, 51
  Insanity 16, 51
  Prioritization 2, 9, 10
  Sweet Spot 8
Philosophical Foundation 22, 23, 24
  What are your assumptions about project management 24
  What do you want project management to do 22
PMI Knowledge Areas
  Scope Management 66, 67
Prioritization 9
Project 18, 19, 40
Project Management Institute
  PMI 66
Project Managers
  Can only do what your organization supports 48
Project Metrics
  Budget at Completion 83
  Cost Performance Index 84, 85
  Cost Variance 83
  Cumulative Cost Performance Index 85
  DO USE METRICS 87
  Earned Value 83
  Estimate at Completion 85
  Estimate to Complete 84
  Internal Rate of Return (IRR) 85, 87
  Metrics 27, 42, 57, 73, 75, 79, 80, 82, 86, 87
  Return on Investment (ROI) 85, 87
  Schedule Performance Index 84
  Schedule Variance 83
  The Fence Example 81
  Variance at Completion 85
Project Phase
  Execution and Control 12

S

Slogans
  Handles 51
  Insanity 16, 51
  Play the Ball 56
Success Factors 14, 33, 38
  Top 20 Success Factor Table 35, 97

Successful Projects 57
  Characteristics of successful projects 57

T

Triple Constraint 20, 47, 52, 66, 68, 73, 74, 75, 79, 83, 84, 85, 99
  Cost 20, 73, 74, 75, 79, 83, 84, 85
  Iron Triangle 47, 99
  Scope 20, 66
  Three-Legged Stool 21, 47, 99
  Time 3, 11, 20, 52, 54, 68

W

Work Breakdown Structure
  WBS 67

www.ingramcontent.com/pod-product-compliance
Lightning Source LLC
Chambersburg PA
CBHW022024170526
45157CB00003B/1345